INSECT VIROLOGY

Kenneth M. Smith
The Cell Research Institute
The University of Texas
Austin, Texas

1967

ACADEMIC PRESS   New York • London

ACADEMIC PRESS INC.
111 Fifth Avenue, New York, New York 10003

United Kingdom Edition published by
ACADEMIC PRESS INC. (LONDON) LTD.
Berkeley Square House, London W.1

LIBRARY OF CONGRESS CATALOG CARD NUMBER 66–29673

PRINTED IN THE UNITED STATES OF AMERICA

228601

INSECT VIROLOGY

Like us attack'd, their tender bodies know
Mortal mischance, and feel their lot of woe;
Pale sickness shakes alike their little frames;
Whether the tainted air's corrupting streams
Or noxious food the latent poison hold,
Whate'er the cause, infection thins the fold;
Fate triumphs, bodies stain'd with putrid gore
Deform the shelves and fun'rals strow the floor;
—from "The Silkworm," 1527, translated by the Rev.
  Samuel Pullein, 1750. [Courtesy of *The Graduate*
  Journal, **VII** (1), The University of Texas, 1965.]

To W. Gordon Whaley, who made this book possible

PREFACE

The study of viruses, or "virology" as it is now called, has achieved the status of a science in its own right and embraces all viruses attacking any kind of living organism. It is no longer permissible to regard the viruses affecting animals, plants, and bacteria as distinctive agents, and it would need a very experienced observer to differentiate, on morphology alone, between some of the small plant and animal viruses. Nevertheless, in view of the vastness of the subject it is still necessary to deal with it piecemeal, grouped according to the type of host organism. There are several books dealing with the viruses affecting the higher animals, plants, and bacteria, but, so far as I am aware, this is the first book in the English language dealing specifically with the viruses which affect insects.

Although primarily intended for insect virologists it is hoped that the book will be of interest to virologists in other fields, particularly those working with the viruses of higher animals who are too prone to forget that insects are also "animals." It should also be useful to microbiologists and others who wish to know something of this rather neglected aspect of virology.

I have attempted to assemble in one volume all the available information concerning the viruses which attack insects. The different kinds of insect viruses are first described, and after that there is a common plan throughout. The symptomatology and pathology of the diseases, isolation and purification of the viruses, the morphology and chemistry of the virus itself and the host range, when known, are all included. This method of approach also reveals the many gaps in our knowledge of the subject. Other chapters deal with the mode of virus replication, transmission, and latent viral infections. An account is given of a new and rapidly developing technique, the growing of insect tissues in culture. This technique presents great opportunities for studying the virus in the living cell but, as is the case with the plant viruses, insect tissue culture lags far behind the advances made by those studying the viruses of the higher animals.

I have said that we cannot regard viruses as being of an entirely different nature merely because they attack very different types of organisms. This is clearly demonstrated in Chapter XI which deals with the relationships of plant viruses with the insects which transmit them. Some plant viruses are shown to multiply in their insect vectors, and even to cause disease in them; thus, some plant viruses are also animal viruses or at least, to quote Aristotle, "the boundary lines between them are indistinct and doubtful."

The last chapter deals with the use of insect viruses in the biological control of insect pests. Much has been heard of the dangers associated with the use or misuse of chemical insecticides. The virus, however, avoids these dangers since it pinpoints only the insect pest and does not, unlike chemical insecticides, destroy bees, parasites, and other beneficial insects. Moreover, since many viruses are transmitted through the egg, it is possible to infect future generations of insects, something no insecticide can do.

I am grateful to Academic Press for allowing me to reproduce part of the chapter on Plant Virus–Vector Relationships from *Advances in Virus Research,* 1965; and for lending me the blocks of several illustrations from "Insect Pathology," Vol. I, 1963, Chapter 14, and *Virology* 27, 1965.

Figures 1–16, 18, 19, 21, 24, 25, 28, 29, 30, 41 and 42 were all made at the Agricultural Research Council Virus Research Unit, Cambridge, England. I also thank several friends who have supplied me with prints of illustrations from their published work; credit has been given to authors in the illustration legends.

December, 1966

KENNETH M. SMITH

CONTENTS

PREFACE vii

CHAPTER I. *Introduction*

Text 1

References 6

CHAPTER II. *The Various Types of Insect Viruses and the Nuclear Polyhedroses*

I. Different Kinds of Insect Virus Diseases 8
II. The Polyhedroses: Nuclear Type 9
   A. The Polyhedra 10
   B. Symptomatology and Pathology 15
   C. Isolation of the Virus 20
   D. Morphology and Ultrastructure of the Virus Particle 22
   E. Chemistry of the Virus 25
   F. The Nuclear Polyhedrosis of *Tipula paludosa* Meig. 30
   G. A Possible Nuclear Polyhedrosis in *Culex tarsalis* Coquillet 35
   H. The Virus 37
   References 37

CHAPTER III. *The Polyhedroses: Cytoplasmic Type*

   A. The Polyhedra 42
   B. Symptomatology and Pathology 47
   C. Isolation of the Virus 49
   D. Morphology and Ultrastructure of the Virus Particle 52
   E. Chemistry of the Virus 57
   References 57

CHAPTER IV. *The Granuloses*

Introduction 59
   A. The Capsules (Granules) 61
   B. Symptomatology and Pathology 62

C. Isolation of the Virus 66
D. Morphology and Ultrastructure of the Virus 68
E. Chemistry of the Virus 71
References 72

CHAPTER V. *The Noninclusion and Miscellaneous Virus Diseases*

Part I. The Noninclusion Diseases

Introduction 74
A. The *Tipula* Iridescent Virus (TIV) 74
B. *Sericesthis* Iridescent Virus (SIV) 86
C. Acute Bee Paralysis Virus (ABPV) 89
D. Chronic Bee Paralysis Virus (CBPV) 94
E. Sacbrood Virus (SBV) 95
F. Wassersucht Virus of Coleopterous Insects 98
G. A Virus from the Armyworm *Cirphis unipuncta* (Haw.) 102
H. A Virus from *Antheraea eucalypti* Scott 102
I. *Drosophila* σ Virus 104

Part II. Miscellaneous Virus Diseases

A. A Virus from *Melolontha melolontha* (Linn.) 105
B. A Suspected Noninclusion Virus in the European Corn Borer
   Ostrinia nubilalis (Hübner) 107
References 107

CHAPTER VI. *Mode of Replication of Insect Viruses*

Introduction 110
A. Nuclear Polyhedrosis Viruses 111
B. Cytoplasmic Polyhedrosis Viruses 114
C. Granulosis Viruses 115
D. Noninclusion-Body Viruses 118
References 125

CHAPTER VII. *Transmission and Spread of Insect Viruses*

A. Infection *per os* 128
B. Transovarial Transmission 130
C. Artificial Methods of Transmission 133
D. Cross Transmission 135
E. Methods of Spread 140
References 142

CHAPTER VIII. *Latent Viral Infections*

A. Definitions 146
B. State of the Virus in Latent Insect Infections 147
C. Examples of Latent Infections 148
D. Conditions Governing Latency 150
E. Induction of Latent Virus Infections 150
References 153

CHAPTER IX. *Tissue Culture of Insect Viruses*

Introduction 155
A. Techniques and Media 160
B. Results Achieved 168
References 170

CHAPTER X. *Further Aspects of the Relationships between Insects and Viruses*

A. Mixed Infections, Interference and Synergism 172
B. Immunity and Resistance 175
C. Virus Strains and Mutations 179
D. Serology of Insect Viruses 180
E. Artificial Feeding Media 182
F. Staining Methods for Optical Microscopy 188
References 191

CHAPTER XI. *Plant Virus–Insect Vector Relationships*

Introduction 195
A. Mechanical Vectors 196
B. The Vector Relationships of Tobacco Mosaic Virus 197
C. The Problem of Aphid-Virus Relationships 202
D. Biological Transmission 215
E. Discussion 226
References 227

CHAPTER XII. *Viruses and the Biological Control of Insect Pests*

Introduction 231
A. Selection of Viruses 232
B. Variable Factors 232

C. Application of the Virus 234
D. Preparing and Storing the Virus 235
E. Standardization of Virus Preparations 236
F. Some Examples of Virus Control 238
References 238

APPENDIX 240

Arthropoda: Arachnida 240
References 242

AUTHOR INDEX 243

SUBJECT INDEX 250

INSECT VIROLOGY

CHAPTER I

Introduction

Virology, that useful but ugly word of doubtful parentage, comprises the study of viruses, irrespective of the host organism from which the virus comes, just as *bacteriology* comprises the study of bacteria. It is necessary to emphasize this point because for many years there was a tendency, especially among workers studying the viruses affecting the higher animals, to regard the virus diseases of bacteria, plants, insects, and higher animals as caused by four separate and distinct types of disease agents. The time has now long gone by when it could be stated that "plant viruses are of an entirely different nature from that of the animal viruses." A virus is a virus regardless of its derivation and, as we shall see later in this book, the multiplication of a plant virus inside an animal, the insect vector, emphasizes the artificiality of host barriers.

It may appear, perhaps, that to write a book on insect viruses is helping to perpetuate these artificial barriers but a moment's thought will show that the immensity of the subject necessitates a piecemeal approach.

More than 300 separate viruses affecting plants alone have been described, and probably a similar number can be recorded from the higher animals, including man. There is no reason to believe that the insect viruses are any fewer in number. There are several books dealing with the viruses of bacteria, plants, and vertebrate animals so that one describing the insect viruses may not be out of place. This branch of virology has been much neglected and intensive study of the viruses infecting insects is of comparatively recent date; in consequence, our knowledge of this group lags behind that of the others.

Although the emergence of virology as a scientific discipline in its own right is so recent, it is possible to look far back in history for events

1

which foreshadowed the development of our subject. So far as plant viruses are concerned the first record in the literature of which we have knowledge is a description published in 1576 by Charles Lécluse or Carolus Clusius of a variegation in the color of tulips which is now called "a color break" and is recognized as being due to an aphid-transmitted virus of the mosaic type. "Broken" tulips are figured in "Theatrum Florae," published in 1662; these illustrations have been identified as the work of the painter Daniel Rabel. A somewhat later account published in "Traité des Tulipes" about 1670 contains the first suggestion that the variegation in the flower color might be due to a disease. Although, of course, at that time there was no conception of the true nature of the causative agent, it was known that the condition was infectious and could be transmitted to "unbroken" tulips by means of bulb grafts.

The first reference to what is now known to be a virus disease of an insect is even earlier. The so-called jaundice disease of the silkworm is described in a poem by Vida, published in 1527 (see page ii).

Scientific proof of the existence of the "ultramicroscopic filter-passing viruses," to use terms now obsolete, was first given in 1892 by a Russian botanist, Dimitri Ivanovsky. He showed that the agent causing mosaic disease of the tobacco plant would pass through a Pasteur-Chamberland filter candle, which removed all bacteria, and would still cause the disease in healthy tobacco. A similar experiment was carried out a few years later in 1898 by two German workers, Loeffler and Frosch, who demonstrated that lymph from an animal infected with foot-and-mouth disease was still infectious after passage through a Berkefeld candle filter. Moreover, by means of a series of dilution experiments they ascertained that the agent was multiplying in the animal host.

Apart from the viruses causing disease in insects with which this book is mainly concerned, viruses are associated with insects in another relationship, one of great importance to man's economy. This is the virus–vector relationship in which the insect plays the essential part of transmitting the virus from one host to another.

Suggestions that insects were somehow connected with the spread of certain diseases of both plants and animals had been made some years before the scientific discovery of a virus.

In 1848 Josiah Nott expressed the opinion that mosquitoes had something to do with the occurrence of yellow fever and about forty years later Carlos Finlay, a local doctor in Cuba, was proclaiming to an unheeding world that there was a connection between mosquitoes and yellow fever. But it was Walter Reed (1902), head of an American Army commission, who finally proved that the yellow fever virus was transmitted by the mosquito *Aëdes aegypti*. The number of animal viruses known to have arthropod vectors is now very large, at least seventy-seven having been described, and they are grouped together under the name of "arboviruses," which does not, however, carry any classificatory significance.

The dependence of plant viruses upon insects for their dissemination is even greater than that of the animal viruses, and the biological relationship between the two in the former case is apparently more fundamental. It has been proved that some animal viruses multiply in their insect vector, notably the virus of equine encephalomyelitis in the mosquito, but they do not seem to have any adverse effect upon the insect. On the other hand certain plant viruses not only multiply in their insect vector but are transmitted transovarially through generation after generation. Furthermore, evidence is accumulating that in some cases the plant virus has a pathological effect on the insect; thus the plant virus is also an insect virus. This subject is dealt with in some detail in Chapter XI.

The following story illustrates the dependence of some plant viruses upon their insect vectors. About 1868 the variegated plant *Abutilon,* probably *A. striatum* var. *Thompsonii,* appeared in Europe and became popular as an ornamental plant. Later, by grafting scions of variegated plants to green shoots of normal plants, it was shown that the variegation was infectious. Because the condition could not be transferred in any other way, it was put in a special category of disease agents and called an "infectious variegation."

Baur (1904) commented on the fact that in Europe variegated and nonvariegated plants were growing side by side without any evidence of transmission. He thought that this fact ruled out the possibility that a living organism was concerned because such a limited capacity for movement was inconsistent with parasitic organisms. Baur further stated that it was important to recognize that there were infectious diseases of which living organisms could not be considered the cause. It seems strange that he should have had to emphasize this fact, which had been amply demonstrated towards the close of the nineteenth century by Ivanovsky in 1892 and Beijerinck in 1898 with tobacco mosaic, and by Loeffler and Frosch in 1898 with foot-and-mouth disease of cattle. However, at that time pathologists were obsessed with the idea that a bacterium, however small, must be the causative agent of every infectious disease, and this idea persisted in some degree right up to 1935 when Stanley isolated the virus of tobacco mosaic.

Quite recently a plant virologist in South America noted a similar variegation on a different but related host plant. Experiment proved that this could be transmitted to *Abutilon* by a specific insect vector, a species of whitefly (Aleurodidae), producing the identical variegation so popular in Europe. This proves, of course, that the condition is not an "infectious variegation" but is due to an insect-transmitted virus of the mosaic type, and the only reason it could not spread in Europe was the absence of this particular species of whitefly, which does not occur there (Orlando and Silberschmidt, 1946).

Insects and other arthropods have therefore a fundamental relationship with viruses from two aspects: (1) as vectors, either mechanical or biological, of plant and animal viruses, and (2) as hosts of their particular type of virus.

It may perhaps be appropriate at this point to try to give the reader some definition of a virus; there have been many attempts to formulate such a definition but the following suggested by Lwoff (1961) seems appropriate: "Viruses represent a specific category of infectious agents which may exist in three states: infectious, proviral and vegetative. The viral infectious particle or *virion* is composed essentially of a condensed genetic material surrounded by a coat or *capsid* formed essentially of proteinic subunits, the *capsomeres*. The virion is, structurally and physiologically, different from any cellular organelle and from any microorganism. It is devoid of metabolism, unable to grow and to undergo binary fission." The writer is not sure how far the statement on inability to grow and divide is applicable to some of the long filamentous viruses such as that of certain forms of influenza or the granulosis virus of insects (see p. 59) which presumably must divide or break up into smaller units.

Two of the larger groups of insect viruses have one fundamental characteristic which sets them apart from all other viruses. This is the occlusion of the virus particle itself inside protein crystals (polyhedra), capsules, and membranes. Once freed of these surrounding envelopes the virus particles resemble morphologically those affecting other types of organisms, and may be small icosahedra or rods. An exception is the long filament associated with the granulosis disease previously mentioned. Insect viruses of the third group are devoid of any intracellular inclusions and are free in the tissues of the insect, similarly to the viruses attacking plants and the higher animals.

The early studies on what are now known to be virus diseases were carried out on the silkworm since this was an insect of great economic importance, and the disease studied was called "jaundice" by the breeders because of the yellow spots which develop on the abdomen of the infected insect in a late stage of the disease. An alternative name was "fatty degeneration" or "grasserie." It is one of what are now classified as the "nuclear polyhedroses," a characteristic of which is the presence in the tissues of huge numbers of many-sided crystals or "polyhedra." The word "polyhedrosis" signifies that type of insect virus disease in which polyhedra are present in the tissues. It was these polyhedra which caught the eye of the early workers, such as Cornalia (1856), Maestri in 1856, and Bolle in 1894, since they were easily visible under the optical microscope. It was Bolle (1894) who first established the protein nature of the polyhedra.

The exact relationship between these polyhedra and the cause of the disease was the subject of extensive studies and controversy for many years. They were variously thought to be reaction products of the disease or a new kind of disease agent. Von Prowazek (1907) demonstrated that material from diseased silkworms was infectious after the polyhedra had all been removed by filtration through many layers of filter paper. This was a significant step towards the realization that viruses were concerned in some insect diseases. Von Prowazek examined his clear filtrate under the microscope and saw that all the polyhedra were removed; this clear fluid was, however, still infectious. He drew from this observation the natural, but erroneous, conclusion that the polyhedra were not the carriers of the virus. It was later suggested by Komárek and Breindl (1924) that the causative agent of the disease might be contained within the polyhedra, since, when these were dissolved in weak alkali, minute objects showing Brownian movement could be observed with the optical microscope by dark-field illumination. This was later confirmed by Bergold (1947) who demonstrated the presence of the virus in the polyhedra by means of the electron microscope.

What von Prowazek did not realize, of course, was that in addition to the virus contained within the polyhedra, there were free virus particles present in his clear filtrate after the removal of the polyhedra, which obviously were invisible under the optical microscope, and those particles produced the disease when introduced into healthy silkworms.

It has been mentioned previously that the viruses affecting insects have received very little attention until recent years and this is a pity because they are very suitable for fundamental study. Insects can be grown in large numbers readily and cheaply, *Galleria mellonella* the wax moth, for example is easily cultured in an artificial medium and with the development of synthetic foodstuffs some lepidopterous and other larvae can be grown in large numbers without the labor of growing plants for food. Recent developments in tissue culture as pioneered by Grace (1962) in Australia and others, offer promise of great insight into the modes of virus replication. Furthermore, viruses can be isolated from insects and purified just as easily as from plants and the higher animals.

Another aspect of insect virology is worth comment here; much has been heard lately of the possible harmful effects of the long-continued use of chemical insecticides upon the fauna of the countryside, including beneficial insects such as bees, and the contamination of foodstuffs.

The use of insect viruses as one aspect of the microbial control of insect pests, which was suggested nearly 60 years ago by von Prowazek, is dealt with in the last chapter of this book. The nuclear polyhedrosis viruses, which are the most suitable for this purpose, can be easily prepared in large quantities at low cost; they can be applied in water or as a dust and they are nontoxic. One great advantage is that they pinpoint the insect pest and leave the beneficial insects unharmed. Furthermore, in many cases these viruses are passed transovarially to the offspring and so it is possible to destroy generations of insects yet unborn, a thing no chemical insecticide can do.

REFERENCES

Baur, E. (1904). *Ber. Deut. Botan. Ges.* **22**, 453; see *Phytopathol. Classics* **7**, 55 (1942).

Beijerinck, M. W. (1898). Ueber ein contagium vivum fluidum als Ursache der Fleckenkrankheit der Tabaksblätter. *Verhandel. Koninkl. Ned. Akad. Wetenschap.* **65**, 3; see *Phytopathol. Classics* **7**, 33 (1942).

Bergold, G. H. (1947). Die Isolierung des Polyeder-Virus und die Natur der Polyeder. Z. *Naturforsch.* **2b**, 122–143.

Bolle, J. (1894). *Jahrb. Seidenbau-Versuchsstat. Gorz* p. 112.

Cornalia, E. (1856). Monografia del bombice del gelso. *Rend. Ist. Lombardo Sci. Lettere, Mem. I*, pp. 348–351.

DATE : 8/2

READER NUMBER : 9180 8065

BOOK NUMBERS
0004 11589

RENEWED UNTIL :

$$
\begin{array}{r}
19\cancel{8}3 \\
1 \ 9 \ \big| \ 6 \ 7 \\
\hline
2 \ 6 \\
+ \ 6 \ 0 \\
\hline
8 \ 6
\end{array}
$$

Grace, T. D. C. (1962). Establishment of four strains of cells from insect tissues grown *in vitro. Nature* 195, 788–789.

Ivanovsky, D. (1892). *St. Petersburg Acad. Imp. Sci. Bull.* 35, 67; see *Phytopathol. Classics* 7, 27 (1942).

Komárek, J., and Breindle, V. (1924). Die Wippelkrankheit-der Nonne und der Erreger derselben. *Z. Angew. Entomol.* 10, 99–162.

Loeffler, F., and Frosch, F. (1898). *Centr. Bakteriol., Parasitenk., Abt. I* 23, 371–391.

Lwoff, A. (1961). The dynamics of viral functions. *Proc. Roy. Soc.* B154, 1–20.

Maestri, A. (1856). Del giallume. *In* "Frammenti anatomici, fisiologici e patologici sul baco da seta," pp. 117–120. Fusi, Pavia.

Orlando, A., and Silberschmidt, K. (1946). *Arquiv. Institute Biol. (Sâo Paulo)* 17, 1.

Reed, W. (1902). Recent researches concerning the etiology, propagation, and prevention of yellow fever by the United States Army Commission. *J. Hyg.* II, 101–119.

Stanley, W. M. (1935). Isolation of a crystalline protein possessing the properties of tobacco-mosaic virus. *Science* 81, 644–645.

von Prowazek, S. (1907). Chlamydozoa *II* Gelbsucht des Seidenraupen. *Arch. Protistenk.* 10, 358–364.

CHAPTER II

The Various Types of Insect Viruses
and the Nuclear Polyhedroses

I. Different Kinds of Insect Virus Diseases

Before describing in detail the viruses attacking insects and the diseases they cause, it will be well to consider the different kinds of viruses and their distribution throughout the insect kingdom.

In the vast majority of insect virus diseases only the larval stage is attacked; there are exceptions to this rule which will be described later. This does not mean that the adult insect, at least in the Lepidoptera, is immune to infection but in most cases the resistance is strong. Caterpillars infected late in their developmental stages may show no sign of disease but occasionally give rise to adults which die shortly after emergence. This is particularly true of the polyhedroses; it is necessary, however, to differentiate between an *active* viral infection of the adult insect and a *latent* infection which usually involves the transovarial spread of the virus.

We have referred in the previous chapter to the occlusion of the virus particles in protein crystals and other enveloping structures. On this basis we can divide the insect viruses into two types, the inclusion and the noninclusion diseases, the former containing, so far as present knowledge goes, the great majority of the insect viruses.

The inclusion diseases consist of two large groups: (1) the *polyhedroses* and (2) the *granuloses*. In the first-named group many hundreds of virus particles are occluded in protein crystals; in the second, a single rod-shaped virus particle (rarely two) is contained in a minute crystal or "granule," sometimes known as a "capsule." The polyhedroses are subdivided into nuclear and cytoplasmic diseases.

8

In the present state of our knowledge of the insect viruses comparatively few, less than a dozen, noninclusion viruses are known. What may be an insect virus which does not fit into any of these categories has been recently described as affecting the larva of the cockchafer, *Melolontha melolontha* (Vago, 1963).

If a survey is made of the distribution of virus diseases throughout the insect kingdom, it is at once apparent that there are numerous orders of insects from which no virus has yet been recorded. It is probable that this uneven distribution is more apparent than real and is merely a reflection of our lack of knowledge of the subject. Nevertheless, no virus infections have been observed so far in the Orthoptera or the Hemiptera, and very few in the Diptera, Coleoptera, and Neuroptera. On the other hand, viruses have been isolated from certain species of spider, or red mites which belong to the class Arachnida and are outside the class of insects altogether. These are briefly dealt with in the Appendix. It is likely therefore that in due course virus diseases will be found affecting members of all the insect orders.

II. The Polyhedroses: Nuclear Type

In this type of disease, as the name implies, the virus multiplies in the cell *nucleus;* the tissues attacked are the epidermis, fat body, blood cells, and tracheae, rarely the silk glands (K. M. Smith and Xeros, 1953a; Aruga *et al.,* 1963). There are two cases known in which the virus appears to multiply in the nuclei of the epithelial cells of the midgut, in a sawfly larva *Gilpinia hercyniae* (Htg.) (Balch and Bird, 1944) and in a lepidopterous larva *Plusia chalcytes* (Esp.) (Laudeho and Amargier, 1963).

The virus particles of the nuclear polyhedroses appear always to be rod-shaped; this is in sharp contrast to the shape of the virus particles of the cytoplasmic polyhedroses which are near-spherical (icosahedra).

The great majority of the nuclear polyhedroses occur in the larvae of the Lepidoptera, especially of the Heterocera. Similar diseases have also been described in hymenopterous insects, in larvae of three or four species of sawflies belonging to the genera *Diprion, Neodiprion,* and *Gilpinia.*

There is only one nuclear polyhedrosis definitely recorded from the Diptera. This attacks the larva of the crane fly, *Tipula paludosa* (Meigen); the virus causing this disease is of unusual interest and is discussed in some detail later in this chapter. A preliminary report describes a possible polyhedrosis from larvae of the mosquito *Culex tarsalis* Coq., but more information on this is needed (Kellen *et al.*, 1963).

Finally, certain species of Neuroptera, e.g., *Chrysopa perla* L., are susceptible to a nuclear polyhedrosis. It is not certain, however, whether this occurs naturally or is only an experimental infection induced under laboratory conditions. The subject of cross transmission is dealt with further in Chapter VIII.

A. The Polyhedra

a. SIZE AND STRUCTURE

The nuclear polyhedra vary considerably, both in size and shape. This variation occurs both in different insects and in the same insect, but not as a rule in the same cell where the polyhedra tend to be the same size. Bergold (1963) points out that in the silkworm, *Bombyx mori*, the prevailing types of polyhedra are dodecahedra, whereas those of *Lymantria monacha* consist mostly of tetrahedra. The polyhedra from *Porthetria dispar* are irregular in shape. The diameter of the polyhedra varies from 0.5 to 15 μ, according to the species. In the case of a nuclear polyhedrosis of *Barathra brassicae*, the diameters varied from 0.8 to 2.7 μ with an average of 1.8 μ (Ponsen and de Jong, 1964).

In the nuclear polyhedroses of some species the polyhedra have an extremely characteristic appearance. In the larvae of the scarlet tiger moth *Panaxia dominula* for example, the polyhedra are rectangular (K. M. Smith, 1955). Similarly in *Tipula paludosa* the polyhedra are usually crescent-shaped, rather like an orange segment.

Gershenson (1959, 1960) considers that it is the virus that controls the shape of the polyhedra rather than the host cell, and he isolated a strain of virus from *Antheraea pernyi* Guérin-Ménéville which induced the formation of a hexagonal polyhedron instead of the more usual tetragon-tritetrahedral shape. This discovery is of some importance because the shape of the polyhedra can be used as a marker to pinpoint a virus strain and thus becomes a useful tool in the study of virus interference and mixed virus infections (see Chapter X).

Examination of ultrathin sections of nuclear polyhedra at very high magnification under the electron microscope reveals very regular dot and line patterns. Bergold (1963) has made a study of the crystalline lattice in the polyhedra of various species by means of X-rays and electron microscopy of very thin sections calculated to be only 100 Å thick and his conclusions are given here. The crystalline lattice has a very high degree of regularity without dislocations, the protein molecules being spheres with a diameter of about 90 Å (Hall, 1960), but this diameter may vary from 65 to 90 Å according to the insect species from which the polyhedra have been obtained. Between the rows of molecules angles of 90° and 120° could readily be observed. All the observed dot and line patterns can be explained with the aid of light and X-ray micrographs of molecule models arranged in a cubic system but cut at different angles.

The arrangement of inclusion-body protein molecules in a face-centered cubic system indicates that there are six selective points of special attraction on the surface of each molecule, which lead to the cubic system. Without these selective points of attraction, one would expect a close-packed hexagonal arrangement which has not been observed in any inclusion body.

In the general run of nuclear polyhedra the virus particles are distributed irregularly and at random, although this does not appear to be the case with the nuclear polyhedra of *Tipula paludosa* (see p. 30). Bergold points out the unexplained fact that the virus particles do not disturb the crystalline lattice (Fig. 1) or act as crystallization centers, though the latter may occur in the disease mentioned above (see Fig. 8).

No components other than virus particles have ever been observed within the polyhedra.

b. PHYSICOCHEMICAL PROPERTIES

The nuclear polyhedra are insoluble in water but dissolve readily in aqueous solutions of NaOH, KOH, NH_3, H_2SO_4, and CH_3COOH. This does not apply, however, to the nuclear polyhedra of *Tipula paludosa* which are dealt with separately later in this chapter.

The polyhedra from different nuclear polyhedroses differ greatly in their resistance to alkali treatment, and this has to be taken into account when isolating the virus from them. Some of the most resistant polyhedra are found in the Australian pasture caterpillar, *Pterolocera amplicornis* Walker; they require 60 minutes of exposure to 4% sodium carbonate at a temperature of 56°C to dissolve them completely (Day *et al.*, 1953).

The effect of sodium hypochlorite upon nuclear polyhedra has also

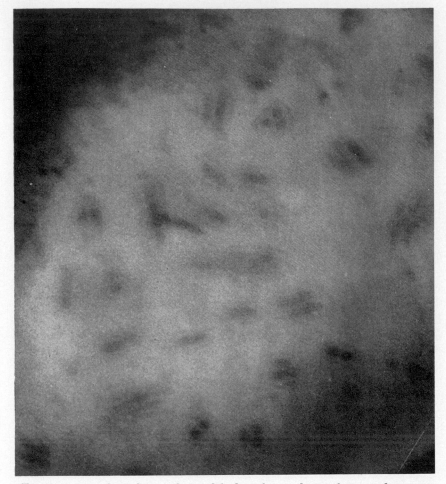

Fig. 1. Section through a nuclear polyhedron from a larva of *Hemerobius stigma* Steph. (Neuroptera). Note that the virus particles do not disturb the crystalline lattice. Magnification, × 200,000.

been studied, the dissolution of the polyhedra being related to concentration of hypochlorite and length of exposure.

In the case of a nuclear polyhedrosis of the cabbage looper, the complete dissolution of polyhedra at 2, 12, and 22 minutes with 0.5, 0.1, and 0.05% hypochlorite, respectively, indicated that the rate of dissolution varied directly with the concentration (Ignoffo and Dutky, 1963).

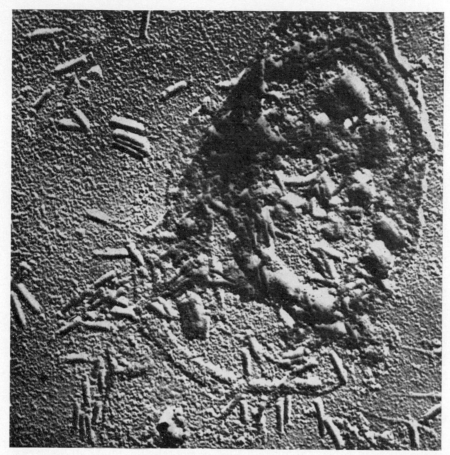

Fig. 2. Nuclear polyhedron dissolved in weak alkali. Note the liberated virus rods and the apparent membrane. Magnification, $\times$ 50,000.

When the nuclear polyhedra are dissolved in weak alkali what appears to be a membrane is left behind, usually enclosing the virus rods which had been occluded in the crystal (Fig. 2). There is some doubt as to the nature of this membrane; it is possibly an artifact in the sense that it is the hardened outer layer of the crystal and Wyatt (1950) has shown that it is of the same composition as the polyhedron itself. It is significant that in thin sections of polyhedra this "membrane" is not visible; this suggests that it is in fact an artifact.

According to Bergold (1963) the polyhedra are resistant to bacterial putrefaction, but Benz (1963) considers that this "membrane" plays a part in protecting the virus particles from the effects of such putrefaction.

When centrifuged out of suspension and washed and dried, the nuclear polyhedra remain apparently unchanged for long periods. Steinhaus (1960) records that in the case of *Bombyx mori*, the silkworm, polyhedra were still infectious for silkworms after 20 years' storage, mostly at refrigerator temperature. Also the solubility of *B. mori* polyhedra in Na_2CO_3 does not change after storage in a desiccator over calcium chloride for 37 years (Aizawa, 1954).

The isoelectric point of *B. mori* polyhedra is pH 5.2 (Tarasevich, 1945), for *Porthetria dispar* pH 5.7, and for *Lymantria monacha* between pH 5.6 and 5.3. The polyhedral proteins are completely insoluble at their isoelectric points. (Bergold and Schramm, 1942).

A clear yellow solution of polyhedral protein can be obtained by dissolving the polyhedra in weak solutions of alkali (0.005 M to 0.03 M Na_2CO_3 + 0.05 M NaCl for 1 to 3 hours). The virus particles are then centrifuged off at 10,000 to 12,000 g; a further ultracentrifugation for 30 minutes at 25,000 g removes all remaining virus particles and their membranes. The polyhedral proteins can then be precipitated by lowering the pH with HCl or by dialysis against distilled water. Under certain salt and pH conditions the polyhedron protein solutions sediment homogeneously on the ultracentrifuge, the sedimentation constant being about 12.5 (Bergold, 1963).

c. CHEMICAL COMPOSITION

Bolle (1874, 1894) was the first to analyze nuclear polyhedra; he found them to consist of protein and to contain no lipids. It must be remembered that any analysis of the whole polyhedra is an analysis of both polyhedral protein and the virus particles, the virus being about 5% of the whole polyhedral crystal. Chemical analyses together with sedimentation and diffusion measurements have revealed that the polyhedra of *Porthetria dispar* and *Bombyx mori* consist of 95% protein with molecular weights of 276,000 and 378,000, respectively (Morgan *et al.*, 1955).

Wellington (1951, 1954) has studied the amino acid composition of a number of insect viruses and their inclusion-body proteins. She found a strikingly greater content of arginine and serine in the virus than in the polyhedra but the latter had more lysine and tyrosine. However, Kawase

(1964) found more arginine in the inclusion body than in the virus. He agrees with Wellington in finding more tyrosine in the polyhedra.

Faulkner (1962) found that the nuclear polyhedra of *B. mori* contained RNA, the purpose of which is not clear. Aizawa and Iida (1963) state that those nuclear polyhedra always contain not only RNA but DNA as well. The RNA content seemed to vary with the silkworm strains under study.

B. Symptomatology and Pathology

The external symptoms of insect virus diseases are governed to a large extent by the tissues affected; this is particularly true as regards the nuclear and cytoplasmic polyhedroses. In the former the skin, blood cells (Fig. 3), fat body, and tracheae are primarily attacked with occa-

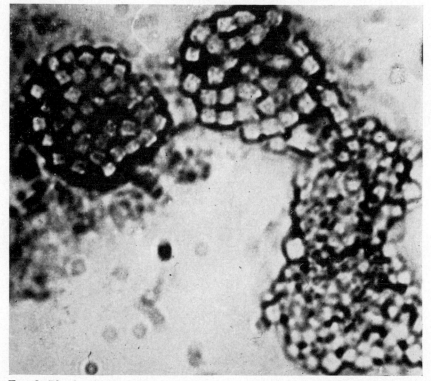

Fig. 3. Blood cells from *Panaxia dominula* (Arctiidae) in a late stage of a nuclear polyhedrosis. Note the rupture of the blood cells and the cubic nature of the polyhedra. Magnification, × 1200.

sional spread to the silk glands (Aruga *et al.*, 1963). The incubation period varies somewhat; it may be as short as 4 days or as long as 3 weeks and the larva may show no symptoms during a large part of this period. As rule, in the early stages of active disease the caterpillar becomes sluggish and refuses to eat. The first definite external sign can be noticed in the skin; in some larvae, notably those of the Vanessids, *Vanessa io* and *V. urticae*, the skin may take on an oily appearance. In the silkworm yellow patches frequently develop, from which the old name of "jaundice" is derived. As the disease progresses the skin becomes exceedingly fragile until it ruptures, liberating the now liquefied body contents consisting in the main of millions of polyhedra. This breakdown of the skin and the subsequent liberation of the body contents are characteristic of the nuclear polyhedroses and the granuloses and differentiate these two diseases sharply from the cytoplasmic polyhedroses. A characteristic of the nuclear polyhedroses is the tendency of affected caterpillars, in a late stage of the disease, to seek the highest point available, whether it be the lid of the cage or the top of a tree, and thence to hang head downwards (Fig. 4). In forests in Germany, where mass infestations of nun moth *Lymantria monacha* Linn. and gypsy moth *Porthetria dispar* Linn. caterpillars are common, the phenomenon is called the Wipfelkrankheit or "tree-top disease."

As the name of the disease implies the cell nucleus is the site of virus multiplication and it is there that the polyhedra are formed. In the silkworm, at an early stage of infection, aggregates of chromatin and very small granules showing strong Brownian movement appear in the nuclear ring zone. These have been called "propolyhedra" and they measure 0.2–0.4 μ in diameter (K. M. Smith and Xeros, 1953b). These granules are early stages in the development of polyhedra which grow around a dense central mass.

Hypertrophy of the cell nucleus is the first sign of infection in the nuclear polyhedroses. Sections made through the cells in the hypodermis of the larva of *Abraxas grossulariata* and through the blood cells of the larva of *Panaxia dominula*, a species of tiger moth, at an early stage of the disease, show the increase in size of the cell nucleus as polyhedra formation begins. As the polyhedra grow, the nucleus completely fills the cell, which eventually bursts, liberating the polyhedra into the hoemocele.

Morris (1962a) describes the histopathology in a nuclear polyhedrosis of the western oak looper, *Lambdina fiscellaria somniaria* Hulst., which

Fig. 4. Larvae of *Vanessa urticae*, the small tortoise-shell butterfly, in a late stage of a nuclear polyhedrosis. Note the characteristic manner in which the larvae hang head downwards.

17

is fairly typical for this type of disease. Following hypertrophy of the fat-body nuclei, granularity of the nucleus increased between the first and third days after infection. Central condensation of the chromatin, which apparently is what Xeros (1956) described as virogenic stroma, then starts, together with the formation of ring zones in the fat-body cells and a few cells of the tracheal epithelium and hypodermis. On the fifth day following infection, small inclusion bodies were visible in the cells of fat body, hypodermis, and tracheal epithelium; these were probably developing polyhedra, possibly the propolyhedra of K. M. Smith and Xeros (1953b). Some blood cells showed "ring zones," and others showed small particulate inclusions peripherally located in the nuclei.

The sequence of events in the development of a nuclear polyhedrosis is fairly consistent in caterpillars of the Lepidoptera. Two or three days after infection, signs of polyhedral formation and coagulation of the chromatin can be detected in the cell nuclei. In the case of a nuclear polyhedrosis of a geometrid moth, *Erannis tiliaria* (Harris), the cells become hypertrophied and die 4 or 5 days after infection. If larvae succeed in reaching the fourth or fifth instar, the polyhedra appear in the nuclei of fat-body cells and in cells of the hypodermis. The greatest quantity of polyhedra is apparently formed 10–15 hours before death (Smirnoff, 1962).

Benz (1963) has made a comprehensive study of a nuclear polyhedrosis affecting the alpine tent caterpillar, *Malacosoma alpicola* (Staudinger) and much of the following information is derived from his paper.

The preliminary signs of infection are characteristic of this type of disease: sluggishness, exudation of fluid, and finally rupture of the skin and liberation of the liquefied body contents. Examination of caterpillars in different stages of the disease suggests that the polyhedra develop sequentially in the following tissues: (1) fat body, (2) hypodermis, (3) tracheal matrix, (4) muscular sheath, (5) nerve sheath, (6) muscles, (7) ganglia, and (8) pericardial cells.

The first cytopathological changes are, as previously remarked, the swelling of the infected nuclei and their nucleoli. By means of metachromatic toluidine blue staining (Pelling, 1959) a substantial increase of nuclear ribonucleic acid (RNA) and later of cytoplasmic RNA is revealed. These changes are first seen in parts of the fat body situated next to the midgut and later in the fat body below the hypodermis and in the hypodermis.

After about 4 or 5 days the nuclei of the fat body commence to swell, and combined with this symptom is a slight increase of Feulgen-positive material. Next the chromatin forms a loose network which is partially replaced by a primarily Feulgen-negative fine-meshed virogenic stroma as described by Xeros (1955, 1956). During their growth the stromata become increasingly Feulgen-positive. In the nuclei of the cells of the fat body and of the hemocytes, polyhedra formation starts in the slightly Feulgen-positive ring zone. The virogenic stroma may persist in the nuclei for a time but gradually disintegrates as polyhedra formation continues. Some active nucleoli may persist until the nuclei are nearly filled with polyhedra. This indicates that protein synthesis continues until polyhedron formation ceases. While the nuclear RNA disappears completely, the concentration of cytoplasmic RNA remains higher than it is in corresponding healthy cells. These primary reactions of the infected cells may be interpreted as an abortive defense reaction.

Sequential changes in DNA, glycogen, and protein indices in the fat body of the western oak looper following infection with a nuclear polyhedrosis have been studied by Morris (1962b). The results indicate that DNA and nuclear protein increased progressively up to a point just prior to polyhedron formation, i.e. about a week after infection. Beyond this point there was apparently a breakdown of DNA accompanied by a further increase in nuclear protein synthesis. Glycogen was drastically reduced to mere traces between the fourth and fifth days following infection but this effect may be in part due to starvation, since the larvae cease to feed.

It has been thought until recently that the endodermal cells of the midgut were never affected in the nuclear polyhedroses of lepidopterous larvae, but this now seems not to be so. An unusual disease of this type has recently been described in the larva of a moth *Plusia chalcytes* (Esp.). Polyhedra were observed in the nuclei of the midgut cells in addition to those of the tissues normally affected in nuclear polyhedroses (Laudeho and Amargier, 1963).

These diseases have so far been described as they affect the larvae of Lepidoptera but the larvae of sawflies, Hymenoptera, are also susceptible to similar infections. Three species have been recorded as infected with nuclear polyhedroses; these are the European spruce sawfly, *Diprion hercyniae* (Hartig), the European pine sawfly, *Neodiprion sertifer* (Geoffroy), and the jackpine sawfly, *Neodiprion pratti banksianae* Rohiver.

The external symptoms of the affected sawfly larvae are similar to those of lepidopterous larvae. There may be a faint yellowish discoloration on the third to fifth abdominal segments, and a milky white fluid may be emitted from the mouth while a dark brown fluid is often exuded from the anus. In the final stage of the disease the skin is ruptured and the liquid body contents are liberated; the dead larvae usually hang head downwards from the branches of the pine trees.

The main difference, however, between the nuclear polyhedroses of lepidopterous and sawfly larvae lies in their histopathology. In the sawfly larvae the polyhedra occur only in the nuclei of the digestive cells of the midgut epithelium, instead of in the nuclei of cells of the blood, fat body, and tracheae (Balch and Bird, 1944; Bird and Whalen, 1953). The fact that the midgut becomes milky because of the presence of the polyhedra may cause confusion in diagnosing this disease because, as we shall see later, the midgut is the site of multiplication of the cytoplasmic polyhedroses. However, the presence of the polyhedra in the *nuclei* should be sufficient to differentiate the two since the polyhedra of the cytoplasmic polyhedroses are found exclusively in the cell cytoplasm.

C. Isolation of the Virus

The isolation and purification of the viruses of the nuclear polyhedroses must be accomplished in two steps. The first step, of course, is the isolation of the polyhedra; this is fairly simple since they are completely insoluble in water. Infected larvae in a late stage of the disease are cut up and suspended in water in an Erlenmeyer flask or similar container. They can then be left for several days or longer at room temperature to allow the bodies to putrefy and disintegrate; the polyhedra will gradually settle to the bottom of the container as a white layer. The supernatant fluid containing the larval debris is then decanted and the polyhedra are suspended in water; further purification by alternate cycles of low- and high-speed centrifugation yields a white preparation of polyhedra with little impurity. The process of polyhedra extraction can be speeded up by grinding the dead larvae in a mortar and washing the resulting paste through cheesecloth in a filter funnel. A certain proportion of polyhedra is lost by this method, probably by adsorption to the host tissues during the grinding-up process.

Fluorocarbon has been used by Bergold (1959b) for further separation of the polyhedra; these are suspended in a fluorocarbon–water mixture (about 1 : 1) and shaken well. The mixture is then centrifuged briefly and the whitish supernatant, which should still contain the polyhedra, can be used or shaken again with fluorocarbon. The final product should be pure white and show no impurity under the optical microscope; polyhedra prepared in this manner can be stored dry for several years.

The second step is the extraction of the virus rods from the purified polyhedra, and this is done by the careful dissolution of the polyhedra to liberate the occluded virus. This is best accomplished by the use of weak sodium carbonate, the strength of which must be determined according to the species of insect and its associated virus. The concentration of Na_2CO_3 and the duration of application are critical since it is only too easy to dissolve the virus along with the polyhedra if the alkalinity of the solution is too high.

The following method was used by Bergold (1947) and is still generally applicable to nuclear polyhedroses with the exception of that which attacks the larvae of *Tipula paludosa* Meig. (Diptera). This interesting disease is dealt with separately at the end of this chapter.

Five milligrams of polyhedra is used for each milliliter of a solution of 0.004–0.03 M Na_2CO_3 + 0.5 M NaCl. The polyhedra should dissolve at room temperature in about 1–2 hours. During this time the milky suspension becomes opaque; it is centrifuged for about 5 minutes at 2000–4000 g to sediment insoluble impurities. A brownish pellet indicates that the polyhedra have not been properly purified and a white sediment indicates that not enough alkali was used. The supernatant should be bluish-white and consist of the virus particles suspended in the polyhedra protein solution. This supernatant is next centrifuged for 1 hour at about 10,000 g. The virus particles collect in a bluish-white pellet and the clear yellowish supernatant of polyhedral protein is discarded. The virus pellet is next suspended in an equal volume of CO_2-free distilled water and centrifuged once more at 10,000 g for 1 hour. The clear supernatant is discarded and the bluish-white virus pellet is suspended in one-seventh of the original volume in CO_2-free distilled water, resulting in a bluish-white suspension of pure virus particles.

A purer preparation of polyhedra may be obtained if only the hemolymph of infected larvae is used. Usually it is only necessary to wash and centrifuge the suspension twice to obtain a good preparation. The quantity of polyhedra when isolated by this method will, of course, not be so great as that obtained by crushing up large numbers of infected caterpillars.

D. Morphology and Ultrastructure of the Virus Particle

Having dissolved away the polyhedral crystal and separated its contents, the isolation process is still not complete because the virus particles are enclosed in an outer membrane, sometimes called the "developmental membrane." The actual material of the virus particle, the nucleoprotein itself, is contained within the *intimate* membrane.

There may be more than one virus rod enclosed in the outer membrane; in the early studies made by Komárek and Breindl (1924) on the nuclear polyhedrosis of the nun moth, the fact that the virus particles were in bundles made them large enough to be observed by dark-field illumination in the optical microscope as minute specks of light. So far as our present knowledge goes all the virus particles of the nuclear polyhedroses are rod-shaped; this is in sharp contrast to the situation in the cytoplasmic polyhedroses which will be discussed later.

The number of virus rods contained in a bundle varies according to the virus and the species of insect host. The greatest number seems to be found in affected larvae of *Porthetria dispar* (L.) and *Lymantria monacho* (L.). Bergold (1963) records up to nineteen rods in one bundle in larvae of the latter species. On the other hand the virus rods do not occur in bundles in the larvae of *Tipula paludosa* Meigen. (see Fig. 6).

The actual dimensions of the virus rods, in their outer membrane, vary between 20 and 50 mμ in diameter to between 200 and 400 mμ in length. According to Bergold (1963) the size of the particle after isolation from the polyhedral crystal is only about two-thirds that of the particle still occluded in the polyhedron. This discrepancy is probably due to shrinkage during the technique of preparation for electron microscopy. The measurements of virus rods in a few representative diseases follow. In a nuclear polyhedrosis of a noctuid larva *Orthosia incerta* (Hufnagel) the diameter of the virus rods without membranes, determined from sections of polyhedra, was about 30 mμ and the length about 250 mμ (Ponsen and de Jong, 1964). The virus rods from *P. dispar* measured 18.22 mμ in diameter and averaged 280 mμ in length (Morgan *et al.*, 1956). In the western oak looper, *Lambdina fiscellaria somniaria* Hulst., the virus rods measure from 300–340 mμ by 60–70 mμ with an average of 332 by 62 mμ (Morris, 1962a), and in *Malacosoma alpicola* (Staudinger) they are 35–41 mμ wide and 270–370 mμ long (Benz, 1963).

Bergold (1963) has made a schematic drawing of the virus rod as it occurs in the polyhedral crystal and its relationship to the occluding membranes based on sections of the polyhedra. From the periphery to the center can be observed, first the outer (developmental) membrane, about 75 Å thick, then a space between it and the intimate membrane measuring about 40 Å, followed by a layer of lesser density towards the central part of the virus rod about 60 Å wide, and finally the central, dense corelike column of the virus proper with a diameter of about 300 Å.

Knowledge of the ultrastructure of the virus rod itself is somewhat meager, Day *et al.* (1958) consider there is evidence that the osmium-fixed virus rods sometimes contain an axial concentration of dense material but this is not established. It has to be remembered that, in sections of polyhedra, the osmium fixative is able to reach the virus particle and this undoubtedly has a disintegrating effect upon it.

With the advent of negative "staining" for electron microscopy with phosphotungstic acid (PTA) it has become possible to gain some knowledge of the ultrastructure of viruses. In the breakdown of the virus rods from the nuclear polyhedrosis of *B. mori*, the silkworm, when the polyhedra are treated with weak alkali, the first stage in the liberation of the contents of the intimate membrane is the peeling off of the outer membrane (Fig. 5). The membrane breaks in the center and folds backward, thus forming two spheres still joined in the middle. These finally break apart and are thought to be the same as the spherical subunits which Bergold said were discharged from the intimate membrane (Bergold, 1958a). The intimate membrane is then exposed; in this instance, it measures about 20 Å in thickness and has a slightly different structure at either end. At very high magnification the contents of the intimate membrane give the appearance of a widely spaced helix; these contents are discharged from either end of the intimate membrane and appear to uncoil as they flow out (Fig. 5). This helix is considered to be in part deoxyribonucleic acid (DNA) and differs markedly from Bergold's subunits (K. M. Smith and Hills, 1962).

Although the actual assembly of the parts forming a new rod has not yet been observed, the presence of a helical structure similar to that observed in the rod of tobacco mosaic virus (TMV) strongly suggests a parallel method of assembly. Krieg (1961) has also made a study under the electron microscope of the virus rods from nuclear polyhedroses and considers that they show a helical structure. Based on his former

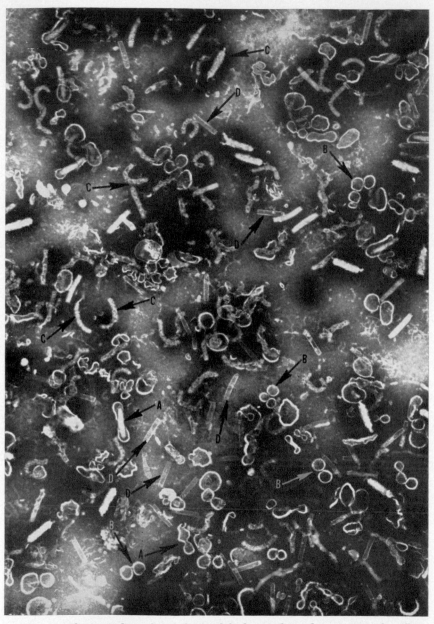

FIG. 5. The virus from the nuclear polyhedrosis of *Bombyx mori* L., the silkworm, treated with weak alkali. Note (A) the capsule breaking in the center and folding backward; (B) the two spheres thus formed; (C) the contents of the intimate membrane which, at high magnification, appear to be helical; in some cases a terminal protrusion is visible; (D) the empty intimate membrane. Magnification, × 30,000.

observations of a central hole in the subunits, he has proposed a model for the rod-shaped viruses of insects similar to that made by Franklin and Klug (1956) for tobacco mosaic virus (TMV).

E. Chemistry of the Virus

Some early work by Breindl and Jírovec (1936) gave the first suggestion that deoxyribonucleic acid (DNA) was associated with the nuclear polyhedrosis virus. They found that the nuclear polyhedra gave a positive Feulgen reaction; we know now that this reaction was due to the occluded virus particles. However, it seems possible that the polyhedra do contain nucleic acid apart from that of the virus rods. Faulkner (1962) found ribonucleic acid (RNA) in the nuclear polyhedra of *Bombyx mori*, the silkworm, and according to Aizawa and Iida (1963) both DNA and RNA are always present in the above polyhedra. However this may be, there is now no doubt that the virus rods of the nuclear polyhedroses do contain DNA and no RNA (Wyatt, 1952a,b). Gratia *et al.* (1945) made the first quantitative examination for nucleic acid, but they also tested the whole polyhedra of *B. mori* and found 0.48% DNA and no RNA.

An analysis of purified virus particles suggested 13% DNA in *B. mori* and 16% in *Porthetria dispar* (Bergold, 1947; Bergold and Pister, 1948). A reinvestigation of very highly purified *B. mori* particles by Bergold and Wellington (1954) revealed the percentage of DNA to be about 7.9% and 0.915% phosphorus, of which, however, only about 87% is found in the DNA.

Krieg (1956) has studied the nucleic acid content of a nuclear polyhedrosis virus from *Aporia crataegi* (Linn.), in which he found 9% DNA but no RNA.

The amino acid content of the nuclear virus protein has been studied by a number of workers. Wellington (1951) made a qualitative analysis by paper chromatography of the amino acids of the nuclear polyhedrosis of *B. mori*. The following amino acids were found in the acid hydrolysates of both the virus and the polyhedral protein: cysteic acid, aspartic acid, glutamic acid, serine, threonine, alanine, tyrosine, methionine sulfone, histidine, lysine, arginine, proline, valine, leucine, isoleucine, phenylalanine, and glycine.

Determinations of tryptophan in the unhydrolyzed samples suggested one difference between the purified viruses and their inclusion-body proteins. The polyhedral proteins had 3.5% of this amino acid whereas the virus failed to respond to the test.

In a later paper Wellington (1954) determined the amino acid content of several nuclear polyhedrosis viruses and their inclusion-body proteins by means of quantitative paper chromatography. She found that whereas the seven viruses analyzed all had a similar pattern of amino acid composition, they differed markedly from the pattern of their respective inclusion-body proteins. In proportion to the total amino acid content,

TABLE I

Amino Acid Composition of Polyhedron Proteins in the Silkworm[a,b,c]

	Hexahedral cytoplasmic inclusions	Icosahedral cytoplasmic inclusions	Nuclear polyhedra
Lysine	6.1	6.3	9.7
Histidine	2.4	2.2	2.2
Ammonia	3.5	4.2	3.6
Arginine	6.2	6.1	6.3
Aspartic acid	13.4	13.6	13.3
Threonine	2.5	3.0	4.3
Serine	7.9	7.6	4.0
Glutamic acid	11.6	12.4	14.3
Proline	3.3	3.2	5.4
Glycine	3.1	3.5	3.3
Alanine	5.1	5.5	3.8
Cystine	0.7	0.9	1.0
Valine	9.3	9.5	6.8
Methionine	1.5	1.3	2.0
Isoleucine	6.7	7.5	6.3
Leucine	7.8	8.6	8.9
Tyrosine	9.8	8.2	8.1
Phenylalanine	6.5	4.8	6.8
Alloisoleucine	0.1	0.1	0.1
Tryptophan[d]	2.3	2.2	2.8
Total	109.8	110.7	113.0
Total calculated as amino acid residue	91.5	91.6	94.7

[a] Data from Kawase (1964).
[b] Material: C122 × N124.
[c] Expressed as gm amino acid per 16 gm nitrogen.
[d] Determined by p-dimethylaminobenzaldehyde method.

the viruses contained a much greater content of arginine and serine, and less lysine and tyrosine than the latter. Kawase (1964) determined the amino acid composition of some polyhedral proteins and viruses of the polyhedroses affecting *B. mori* by means of the Beckman-Spinco Model 120 amino acid analyzer.

Tables I and II are reproduced from Kawase's paper and they show the amino acid content of the polyhedral protein and the viruses, respectively. Three viruses, two cytoplasmic and one nuclear, were examined together with their polyhedral proteins and they showed considerable similarity in the pattern of their amino acid composition. The relative

TABLE II

AMINO ACID CONTENTS OF SILKWORM VIRUSES[a,b,c]

	Hexahedral cytoplasmic inclusions	Icosahedral cytoplasmic inclusions	Nuclear polyhedra
Lysine	5.8	5.7	5.4
Histidine	2.8	2.1	Trace[d]
Ammonia	4.0	4.6	4.0
Arginine	6.5	6.4	5.5
Aspartic acid	12.5	11.4	12.7
Threonine	4.4	4.4	6.2
Serine	6.9	6.9	6.1
Glutamic acid	10.6	10.4	6.2
Proline	4.7	4.8	7.1
Glycine	4.4	5.0	6.4
Alanine	5.6	6.0	7.4
Cystine	Trace	0	0
Valine	5.8	6.6	5.9
Methionine	2.0	1.0	1.5
Isoleucine	5.1	6.1	5.7
Leucine	7.5	8.2	9.2
Tyrosine	6.1	4.3	4.7
Phenylalanine	5.4	6.0	6.2
Alloisoleucine	0	Trace	Trace
Grams of amino acid recovered per 100 gm of virus	65.4	66.4	70.2

[a] Data from Kawase (1964).
[b] Material: Keno × Shunpaku.
[c] Expressed as percentage of total recovered amino acids.
[d] In this sample histidine could not be calculated because of interference by ammonia.

proportion of arginine in the viruses was similar to that of the poly-hedral proteins except that in the nuclear type there was more arginine in the inclusion-body protein than in the virus. The viruses, however, had less tyrosine than the polyhedra, just as reported by Wellington (1954). Furthermore, the viruses had much more threonine, proline, and glycine, and less cystine than the polyhedral protein.

According to Bergold and Wellington (1954), the amino acid com-position of the virus membranes and that of the viruses is quite different; the virus membranes contain more aspartic acid and much less arginine than the virus.

J. D. Smith and Wyatt (1951) and Wyatt (1952b) have investigated the bases of several insect viruses. In the nuclear polyhedrosis of *Lyman-tria dispar*, the gypsy moth, the DNA was shown to contain the purines adenine and guanine and the pyrimidines cytosine and thymine. 5-methylcytosine could not be detected. Chromatograms of the gypsy moth virus hydrolyzed whole, without isolation of its nucleic acid, showed no uracil, thus confirming the absence of RNA from this virus.

The demonstration by Gierer and Schramm (1956) in Germany and by Fraenkel-Conrat (1956) in California that the RNA of tobacco mosaic virus could alone initiate infection in susceptible plants stimulated simi-lar investigations into the infectiousness of the RNA and DNA of other viruses. Gershenson (1956a,b) isolated protein-free DNA and DNA-free protein from the nuclear polyhedra of *Antheraea pernyi*. He found that neither of these was infectious when injected singly. However, by mixing the DNA and the protein for 7 hours before injection, 40% mortality was obtained.

Bergold (1958b, 1959a) isolated the DNA from purified *B. mori* virus using *p*-aminosalicylate and phenol. The infectivity of this preparation, used at the rate of 0.1 mg per larva of *B. mori*, was only 0.0001% that of the corresponding amount of untreated DNA still contained in the intact virus particle.

In a series of experiments Gershenson *et al.* (1963) present evidence that insect cells infected with a virus of DNA type (nuclear poly-hedrosis) produce, instead of the RNA normal for them, a modified type which is capable of inducing a DNA virus when introduced into a sus-ceptible host.

Put briefly, their experiments were as follows. RNA was isolated by a modification of the phenol method from healthy fifth-instar larvae or young pupae of *B. mori*; RNA was similarly isolated from larvae which had been infected with their own nuclear polyhedrosis by injection. The RNA was dissolved in pyrophosphate or phosphate buffer at pH 7.0 and purified by centrifugation (20,000 to 25,000 g) at $0°-4°C$.

Solutions of RNA from healthy or virus-infected larvae or pupae were injected intralymphatically into healthy fifth-instar larvae or pupae of *B. mori*. The results of these experiments were that out of 200 larvae injected with RNA from healthy larvae only one developed a nuclear polyhedrosis, whereas out of 894 larvae injected with RNA from virus-infected larvae 383 (44%) developed a nuclear polyhedrosis.

Gershenson and his co-workers have tested a number of possible alternative explanations. In order to be certain that the effect observed with RNA from infected larvae was not a result of accidental presence of virus in the preparations, the following experiment was carried out. Purified polyhedra from *B. mori* were added to healthy pupae during homogenization and RNA was isolated from the homogenate. This RNA failed to induce nuclear polyhedrosis, thus showing that the virus had been eliminated during the isolation of the RNA. Furthermore, the high-speed centrifugation alone is sufficient to remove the virus.

To ascertain that it was the RNA itself that was the infectious element and not some other component, solutions of the RNA were treated with RNase (ribonuclease); such solutions almost lost their infectivity, giving only 10 diseased larvae out of 191. RNA solutions, on the other hand, treated with DNase gave 37 nuclear polyhedroses in 80 larvae injected. Solutions of RNA from infected insects become practically noninfectious after standing for 24 hours at room temperature, whereas the infectivity of virus suspensions is unaffected under those conditions.

One of the greatest pitfalls in this type of experimental work is the possibility of latent virus infections which may be stimulated into virulence by a variety of conditions (see Chapter VIII). Gershenson considers that this eventuality may be met by the use of a mutant strain of virus with polyhedra of a characteristic shape. Such a strain of virus with hexagonal polyhedra was isolated from a larva of *Antheraea pernyii*. From a number of pupae infected with this virus the RNA was isolated and injected into healthy larvae of the same species. Some of these larvae developed a nuclear polyhedrosis which was characterized by the presence of the hexagonal polyhedra instead of the tetragon-tritetrahedral type of the normal virus. In other words, the shape of the polyhedra is used as a marker to pinpoint the actual virus.

If this work is confirmed it will become evident that genetic information can be transmitted not only from DNA to RNA but also in the reverse direction. This would also confirm the suggestion of Stent (1958) that in some stages of the multiplication of DNA viruses the process is controlled by RNA (and protein) without the participation of DNA.

F. The Nuclear Polyhedrosis of *Tipula paludosa* Meig.

The nuclear polyhedroses of lepidopterous and hymenopterous larvae are fairly similar and adequate descriptions of one or two representative examples have been given. The nuclear polyhedrosis of the larva of the crane fly *Tipula paludosa* Meig., however, differs in many respects from the foregoing and so merits separate consideration.

The disease was first recorded by Rennie (1923) who, however, described it as affecting the fat body. It was rediscovered by K. M. Smith and Xeros (1954) who have made an extensive study of this unusual virus disease.

The external symptoms are entirely different from those of the nuclear polyhedroses of the Lepidoptera and Hymenoptera. As the skin is not attacked, the extreme breakdown characteristic of the other nuclear polyhedroses is not apparent. On the contrary, infected larvae live for a month or more and the larval period may even be prolonged beyond the normal duration, although death eventually ensues. The chief external symptom is a marked pallor, affected larvae being almost white in deep contrast with the gray color of normal larvae. This pallor is probably due to the accumulation of vast quantities of polyhedra within the blood cells; the infection is solely a nuclear polyhedrosis of the blood cells, which seem to increase greatly in number as the disease progresses.

In the enlarged nuclei of the blood cells a chromatic mass forms, and from this mass the chromatic material segregates as several spherical bodies, leaving a large eosinophile body in the nucleus; the polyhedra seem to arise around the periphery and are closely applied to the nuclear membrane (Fig. 6). They differ in appearance and may be rectangular or shaped like an orange segment. They are negatively birefringent and appear to be genuine crystals. Their behavior in the presence of different reagents is most unusual and differentiates them sharply from the other groups of polyhedra. They are resistant to trypsin and to dilute and weak acids and alkalis. In 1 N sodium hydroxide they elongate to six or more times their own length, becoming first biconvex spindles and then elongating into crescents or wormlike shapes. At about three times their

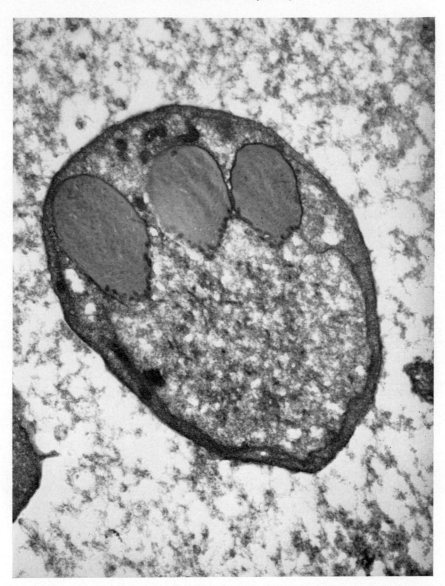

FIG. 6. Section through a blood cell from the larva of the fly, *Tipula paludosa* Meig., infected with its nuclear polyhedrosis. Note the enlarged nucleus and three polyhedra situated on its periphery. The virus particles in the polyhedra appear to have a definite arrangement. Magnification, × 10,000.

normal length this elongation is still completely reversible and in water at pH 5–8 they return to their original size and shape. After such treatment, however, the polyhedra are "activated"; in other words they now respond in a similar manner in ammonia, 1–12% sodium carbonate, and hydrochloric acid, pH 1–4, but not to 1 N hydrochloric acid or 25% sodium carbonate. The elongation or retraction, i.e. return to normal shape, which takes place along the same axis, can be repeated indefinitely in these solutions and takes place as rapidly as the solutions can be alternated.

The polyhedra are not usually dissolved even after half an hour in 1 N sodium hydroxide at 20°C. In a solution of equal parts of 1 N sodium hydroxide and 1 N postassium cyanide, the response is speeded up enormously and the polyhedra elongate even further to reach their maximum in 1½ minutes. They dissolve completely in 2–4 minutes.

The polyhedra give a positive reaction with the xanthoproteic and ninhydrin tests and stain with bromophenol blue after treatment for 15 minutes with 1 N hydrochloric acid at 60°C. Their substance is Feulgen-positive throughout, and each polyhedron has intensely staining granular bands at about ½-μ intervals. Tests for sulfur have been negative.

In such resistant polyhedra as these, it is a difficult matter to observe any virus bodies within; to do this it is necessary to have recourse to thin sectioning of the polyhedra (Fig. 6).

However, studies have been made under the electron microscope of polyhedra subjected to various treatments. In one, the polyhedra were exposed to 1 N sodium hydroxide and washed. The resulting elongated forms appeared semi-opaque with a granular and fibrous structure. In another series the polyhedra were treated with 1 N sodium hydroxide for 4 minutes, then washed and treated with sodium thioglycolate at pH 8.4 for 4 minutes, and finally washed and treated for one minute with 1 N sodium hydroxide. Under this treatment most of the substance of the polyhedral crystal is dissolved leaving behind a lens-shaped ghost. In many polyhedra treated in this fashion were rod-shaped bodies, confirmed as virus rods by subsequent examination in the electron microscope of ultrathin sections of untreated polyhedra (K. M. Smith and Xeros, 1954).

Figure 7 is a photomicrograph of three blood cells of *T. paludosa* in different stages of infection. At the top right is a cell in an early stage with some indications of polyhedral development; at the bottom right the cell is enlarging with the polyhedra mostly formed. At the top left

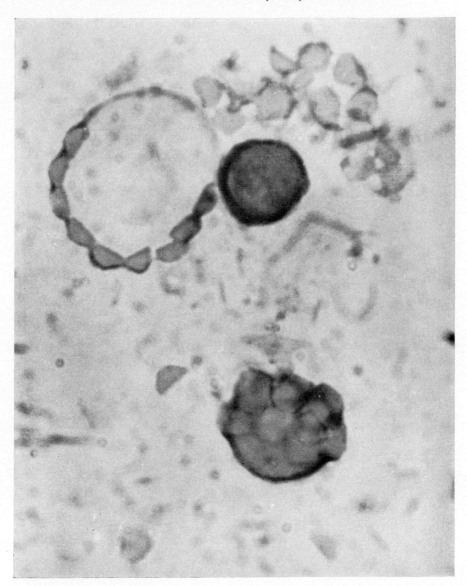

FIG. 7. Three blood cells from a larva of *T. paludosa* in different stages of infection with its nuclear polyhedrosis. For explanation see text. Magnification, × 1500.

the cell has enlarged to bursting point and the majority of the polyhedra have been discharged. Some polyhedra, however, are still in position on the periphery of the nuclear membrane which at this stage completely fills the cell. It will be noticed that the shape of the polyhedra approximates that of a crescent or an orange segment rather than the more usual polyhedron (Fig. 8).

Xeros (1964) has described the phagocytosis of the polyhedra in this disease. K. M. Smith (1956) published an electron micrograph show-

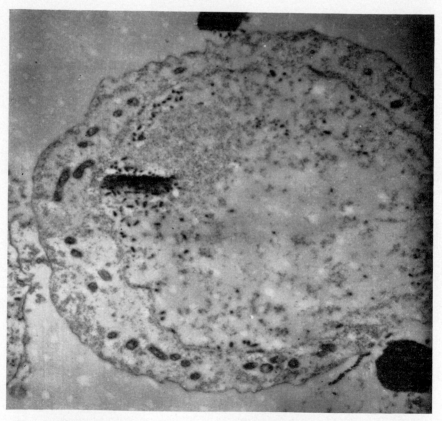

Fig. 8. Electron micrograph of a section through a blood cell from a larva of *T. paludosa* in an early stage of infection with its nuclear polyhedrosis. Note, at left, an apparent center of crystallization of a polyhedron and the migration of the virus rods towards it. Magnification, × 12,500.

ing two bodies about the size of polyhedra and with an appropriate content of virus rods, each of which is enclosed in its outer membrane. The two bodies lack the electron-dense polyhedral matrix and are bounded by a single membrane and not by a thick layer of cytoplasmic membranes. These were present in the cytoplasm alongside a mitochondrion in the cytoplasm and the similarity between the two bodies suggested the possibility of the virus having developed within a mitochondrion. Xeros considers, however, that they are almost certainly phagocytosed virus particles present in the cytoplasm of a nodule cell and secondarily enclosed in a membrane, or phagocytosed polyhedra whose matrix has been dissolved after each was enclosed in a single membrane. A number of peculiar bodies occurring in the nodule cells are thought to be phagocytosed polyhedra (Xeros, 1964). In this connection it is of interest to find that blocking the phagocytes may increase susceptibility to a nuclear polyhedrosis. Larvae of *Galleria mellonella* (L.) pretreated with India ink were more susceptible to infection by virus particles than untreated larvae which were able to tolerate 13 times more virus (Stairs, 1964).

Owing to the difficulty, already described, of freeing the virus particles from the polyhedra in this particular disease, no studies of their chemistry or ultrastructure have so far been possible. Thin sections of polyhedra, however, show that the virus particles are rods enclosed in an outer membrane and appear superficially similar to the virus rods of the other nuclear polyhedroses.

The host range of the *Tipula* nuclear polyhedrosis virus seems to be restricted and with the possible exception of *Tipula oleracea,* there are no records of transmission to other species.

G. A Possible Nuclear Polyhedrosis in *Culex tarsalis* Coquillet

A disease having many characteristics of a nuclear polyhedrosis, and attacking a dipterous larva, has been described by Kellen *et al.* (1963). It affects the larva of the mosquito *Culex tarsalis* Coq. The only previous record of a possible virus disease in a mosquito larva was reported in *Anopheles subpictus* Grassi by Dasgupta and Ray (1954, 1957) who observed Feulgen-positive intranuclear inclusion bodies in the secretory cells of the midgut. The viral nature of these bodies was not, however, definitely established.

a. The Polyhedra

Tetragonal inclusion bodies were observed in nuclei of hypodermal cells and developing adult leg, wing, and antennae buds. Usually only one or two inclusion bodies occupied an infected nucleus; however, masses of about sixteen crystals were observed in some nuclei. Crystals were also present in the cytoplasm of a few cells.

The cuboid crystals usually measured about 2 to 3 μ, but crystals up to 6 μ in diameter were common. Only tetragonal forms were observed. The crystals were insoluble in ethyl alcohol and methyl alcohol but dissolved readily in 1.0 N NaOH at room temperature. A mass of fine residue remained after the crystals dissolved.

b. Symptomatology and Pathology

Under laboratory conditions many larvae became moribund after 3 days. They did not respond normally to disturbances in the water, usually remaining inactive and suspended from the surface film in close groups of about 10 to 50 individuals. When strongly stimulated, the larvae moved sluggishly and had difficulty in regaining attachment to the water surface. The abdomens of affected larvae were abnormally curved or **S**-shaped; the lateral thoracic areas and anterior abdominal segments were inflated with hemolymph. Developing adult structures, such as wings, legs, antennae, and mouthparts, were partially destroyed or malformed and appeared dull white beneath the transparent cuticle. The thoracic cuticle and occasionally certain of the abdominal segments developed hard, shiny black spots about 50 to 100 μ in diameter. Larvae showing these symptoms invariably died in the fourth instar, and usually remained attached to the water surface after death.

c. Transmission

It was found that larvae could be infected *per os,* and larvae so infected survived until the fourth instar.

Successful transmissions were also obtained by exposing first-instar larvae to infected material which had been air-dried at room temperature and stored at $-10°C$ for 12 months; the infected larvae developed characteristic symptoms in the fourth instar. Transmission to fourth-instar larvae has been obtained by intrahemocoelic inoculation with infected hemolymph; these transmissions were effected by abdominal punctures using fine needles.

Similar-appearing inclusion bodies have also been observed in fourth-instar larvae of *Aëdes sierrensis* Ludlow and *Anopheles freeborni* Aitken.

H. The Virus

Preliminary studies with the electron microscope have failed to reveal virus particles after dissolution of the crystals (Kellen *et al.*, 1963).

REFERENCES

Aizawa, K. (1954). Dissolving curve and the virus activity of the polyhedral bodies of *Bombyx mori* L., obtained 37 years ago. *Sanshi Kenkyu* 8, 52–54.

Aizawa, K., and Iida, S. (1963). Nucleic acids extracted from the virus polyhedra of the silkworm, *Bombyx mori* (L.). *J. Insect Pathol.* 5, 344–348.

Aruga, H., Fukuda, S., and Joshitake, N. (1963). Observations on a polyhedrosis virus within the nucleus of the silk gland cell of the silkworm, *Bombyx mori* L. *Nippon Sanshigaku Zasshi* 32, 213–218.

Balch, R. E., and Bird, F. T. (1944). A disease of the European Spruce sawfly, *Gilpinia hercyniae* (Htg.), and its place in natural control. *Sci. Agr.* 25, 65–80.

Benz, G. (1963). A nuclear polyhedrosis of *Malacasoma alpicola* (Staudinger). *J. Insect Pathol.* 5, 215–241.

Bergold, G. H. (1947). Die Isolierung des Polyeder-Virus und die Natur der Polyeder. *Z. Naturforsch.* 2b, 122–143.

Bergold, G. H. (1958a). Viruses of insects. *In* "Handbuch der Virusforschung" (C. Hallauer and K. F. Meyer, eds.), Vol. 4, pp. 60–142. Springer, Vienna.

Bergold, G. H. (1958b). Some topics of insect-virology. *Trans. 1st Intern. Conf. Insect Pathol. & Biol. Control, Praha, 1958* pp. 191–195.

Bergold, G. H. (1959a). Structure and chemistry of insect-viruses. *Proc. 4th Intern. Congr. Biochem. Vienna, 1958* Vol. 7, pp. 95–98. Pergamon Press, Oxford.

Bergold, G. H. (1959b). Purification of insect-virus inclusion bodies with a fluorocarbon. *J. Insect Pathol.* 1, 96–97.

Bergold, G. H. (1963). The nature of nuclear-polyhedrosis viruses. *In* "Insect Pathology" (E. A. Steinhaus, ed.), Vol. 1, p. 417. Academic Press, New York.

Bergold, G. H., and Pister, L. (1948). Zur quantitativen Mikrobestimmung von Desoxy-und Ribonucleinsäure. *Z. Naturforsch.* 3b, 406–410.

Bergold, G. H., and Schramm, G. (1942). Biochemische Characterisierung von Insektenviren. *Biol. Zentr.* 62, 105–118.

Bergold, G. H., and Wellington, E. F. (1954). Isolation and chemical composition of the membranes of an insect virus and their relation to the virus and polyhedral bodies. *J. Bacteriol.* 67, 210–216.

Bird, F. T., and Whalen, M. M. (1953). A virus disease of the European pine sawfly. *Neodiprion sertifer* (Geoffr.) *Can. Entomologist* 85, 433–437.

Bolle, J. (1874). *Jahrb. Seidenbau-Versuchsstat. Gorz* p. 129.

Bolle, J. (1894). *Jahrb. Seidenbau-Versuchsstat. Gorz* p. 112.

Breindl, V., and Jírovec, O. (1936). Polyeder und Polyedervirus im Lichte der Feulgenschen Nucleareaktion. *Vestn. Cesk. Zool. Spolecnosti Praha* 3, 9–11.

Dasgupta, B., and Ray, H. N. (1954). Occurrence of intranuclear inclusion bodies in the larva of *Anopheles subpictus*. *Bull. Calcutta School Trop. Med.*, **2**, 57–58.

Dasgupta, B., and Ray, H. N. (1957). The intranuclear inclusions in the midgut of the larva of *Anopheles subpictus*. *Parasitology* **47**, 194–195.

Day, M. F., Common, I. F. B., Farrant, J. L., and Potter, C. (1953). A polyhedral virus disease of a pasture caterpillar *Pterolocera amplicornis* Walker (Anthelidae). *Australian J. Biol. Sci.* **6**, 574–579.

Day, M. F., Farrant, J. L., and Potter, C. (1958). The structure and development of a polyhedral virus affecting the moth larva, *Pterolocera amplicornis*. *J. Ultrastruct. Res.* **2**, 227–238.

Faulkner, P. (1962). Isolation and analysis of ribonucleic acid from inclusion bodies of the nuclear polyhedrosis of the silkworm. *Virology* **16**, 479–484.

Fraenkel-Conrat, H. (1956). The role of the nucleic acid in the reconstitution of active tobacco mosaic virus. *J. Am. Chem. Soc.* **78**, 882.

Franklin, R. E., and Klug, A. (1956). The nature of the helical groove on the tobacco mosaic virus particle. *Biochim. Biophys. Acta* **19**, 403.

Gershenson, S. M. (1956a). Reconstitution of active polyhedral virus from nucleic acid and protein outside the organism. *Compt. Rend. Acad. Sci. URSS* **5**, 492–493.

Gershenson, S. M. (1956b). Reconstitution of active polyhedral virus from noninfectious protein and nucleic acid. *Bull. Soc. Naturalists Moscow* **61**, 99–101.

Gershenson, S. M. (1959). Mutation of polyhedrosis viruses *Dokl. Akad. Nauk SSSR* **128**, 622–625.

Gershenson, S. M. (1960). A study on a mutant strain of nuclear polyhedral virus of oak silkworm. *Probl. Virol. (USSR) (English Transl.)* **6**, 720–725.

Gershenson, S. M., Kok, I. P., Vitas, K. I., Dobrovolskaya, G. M., and Skuratovskaia, I. N. (1963). Formation of a DNA-containing virus by host RNA. *Proc. 5th Intern. Congr. Biochem., Moscow, 1961* Vol. 9, p. 150. Pergamon Press, Oxford.

Gierer, A., and Schramm, G. (1956). Infectivity of ribonucleic acid from tobacco mosaic virus. *Nature* **177**, 702–703.

Gratia, A., Brachet, J., and Jeenor, R. (1945). Étude histochimique et microchimique des acides nucléiques au cours de la grasserie du ver à soie. *Bull. Acad. Roy. Med. Belg.* **10**, 72–81.

Hall, C. E. (1960). Measurement of globular protein molecules by electron microscopy. *J. Biophys. Biochem. Cytol.* **7**, 613–618.

Ignoffo, C. M., and Dutky, S. R. (1963). The effect of sodium hypochlorite on the viability and infectivity of *Bacillus* and *Beauveria* spores and cabbage looper nuclear polyhedrosis virus. *J. Insect Pathol.* **5**, 422–426.

Kawase, S. (1964). The amino-acid composition of viruses and their polyhedron proteins of the polyhedroses of the silkworm, *B. mori* L. *J. Insect Pathol.* **6**, 156–163.

Kellen, W. R., Clark, T. B., and Lindegren, J. E. (1963). A possible polyhedrosis in *Culex tarsalis* Coq. (Diptera, Culicidae). *J. Insect Pathol.* **5**, 98–103.

Komárek, J., and Breindl, V. (1924). Die Wippelkrankheit der Nonne und der Erreger derselben. *Z. Angew. Entomol.* **10**, 99–162.

Krieg, A. (1956). Über die Nucleinsäuren der Polederviren. *Naturwiss.* **43**, 537.

Krieg, A. (1961). Über den Aufbau und die Vermehrungsmöglichkeiten von Stabchenförmigen Insekten-Viren II. Z. *Naturforsch.* **16b**, 115–117.

Laudeho, Y., and Amargier, A., (1963). Virose a polyèdres nucléaires a localisation inhabituelle chez un lepidoptère. *Rev. Pathol. Vegetale Entomol Agr. France* **42**, 207–210.

Morgan, C., Bergold, G. H., Moore, D. H., and Rose, H. M. (1955). The macromolecular paracrystalline lattice of insect viral polyhedral bodies demonstrated in ultrathin sections examined in the electron microscope. *J. Biophys. Biochem. Cytol.* **1**, 187–190.

Morgan, C., Bergold, G. H., and Rose, H. M. (1956). Use of serial sections to delineate the structure of *Porthetria dispar* virus in the electron microscope. *J. Biophys. Biochem. Cytol.* **2**, 23–28.

Morris, O. N. (1962a). Studies on the causative agent and histopathology of a virus disease of the western oak looper. *J. Insect Pathol.* **4**, 446.

Morris, O. N. (1962b). Progressive histochemical changes in virus-infected fat body of the western oak looper. *J. Insect Pathol.* **4**, 454–464.

Pelling, S. (1959). Chromosomal synthesis of ribonucleic acid as shown by incorporation of uridine labelled with tritium. *Nature* **184**, 655–656.

Ponsen, M. B., and de Jong, D. J. (1964). A nuclear polyhedrosis of *Orthosia incerta* (Hufnagel) Lepidoptera, Noctuidae. *J. Insect Pathol.* **6**, 376–378.

Rennie, J. (1923). Polyhedral disease in *Tipula paludosa* (Meig.) *Proc. Roy. Phys. Soc. Edinburgh* **A20**, 265–267.

Smirnoff, W. A. (1962). A nuclear polyhedrosis of *Erannis tiliaria* (Harris) (Lepidoptera, Geometridae). *J. Insect Pathol.* **4**, 393–400.

Smith, J. D., and Wyatt, G. R. (1951). The composition of some microbial deoxypentose nucleic acids. Biochem. J. **49**, 144–148.

Smith, K. M. (1955). Morphology and development of insect viruses. *Advan. Virus Res.* **3**, 199–220.

Smith, K. M. (1956). The electron microscope in the study of viruses. *Lectures Sci. Basis Med.* **6**, 379.

Smith, K. M., and Hills, G. J. (1962). Ultrastructure and replication of insect viruses. *Proc. 5th Intern. Conf. Electron Microscopy, Philadelphia, 1962 Vol. 2.* Art. V-1. Academic Press, New York.

Smith, K. M., and Xeros, N. (1953a). Cross-inoculation studies with polyhedral viruses. *Symp. Interaction Viruses & Cells, Rome, 1953* pp. 81–96.

Smith, K. M., and Xeros, N. (1953b). Development of virus in the cell nucleus. *Nature* **172**, 670–671.

Smith, K. M., and Xeros, N. (1954). An unusual virus disease of a dipterous larva. *Nature* **173**, 866–867.

Stairs, G. R. (1964). Changes in the susceptibility of *Galleria mellonella* (Linn.) larvae to nuclear-polyhedrosis virus following blockage of the phagocytes with India ink. *J. Insect Pathol.* **6**, 373–386.

Steinhaus, E. A. (1960). The duration of viability and infectivity of certain insect pathogens. *J. Insect Pathol.* **2**, 225–229.

Stent, G. H. (1958). Mating in the reproduction of bacterial viruses. *Advan. Virus Res.* **5**, 95–149.

Tarasevich, L. M. (1945). Determination of the isoelectric point of virus proteins by staining. *Compt. Rend. Acad. Sci. URSS* **47**, 94.

Vago, C. (1963). A new type of insect virus. *J. Insect Pathol.* **5**, 275–276.

Wellington, E. F. (1951). Amino acids of two insect viruses. *Biochim. Biophys. Acta* **7**, 238–243.

Wellington, E. F. (1954). The amino acid composition of some insect viruses and their characteristic inclusion body proteins. *Biochem. J.* **57**, 334–338.

Wyatt, G. R. (1950). Studies on insect viruses and nucleic acids. Ph.D. Thesis, Cambridge University, England.

Wyatt, G. R. (1952a). Specificity in the composition of nucleic acids. *Exptl. Cell Res. Suppl.* **2**, 201–217.

Wyatt, G. R. (1952b). The nucleic acids of some insect viruses. *J. Gen. Physiol.* **36**,

Xeros, N. (1955). Origin of the virus-producing chromatic mass on net of the insect nuclear polyhedroses. *Nature* **175**, 588.

Xeros, N. (1956). The virogenic stroma in nuclear and cytoplasmic polyhedroses. *Nature* **178**, 412–413.

Xeros, N. (1964). Phagocytosis of virus in *Tipula paludosa* Meigen. *J. Insect Pathol.* **6**, 225–236.

CHAPTER III

The Polyhedroses: Cytoplasmic Type

The existence of a separate and distinct type of polyhedrosis virus which was near-spherical instead of the usual rod-shape was first demonstrated by Smith and Wyckoff (1950) in the larvae of *Arctia caja* Linn., and *A. villica* Linn. The differential staining properties of this new type of polyhedra were later investigated (Smith *et al.*, 1953).

Previously other workers had observed polyhedra in the midgut cells (Ishimori, 1934; Lotmar, 1941), but the fact that they were of an entirely different nature from the nuclear polyhedra was not realized. It was later discovered that these polyhedra are formed in the cytoplasm of the cells of the midgut and not in the cell nuclei; it is from this fact that the name "Cytoplasmic Polyhedrosis" was derived (Xeros, 1952; Smith and Xeros, 1953).

This type of virus attacks mainly the larvae of Lepidoptera, both of the Rhopalocera and Heterocera; the cytoplasmic polyhedroses are as numerous as the nuclear polyhedroses, and may even exceed them in number. No fewer than eighty-one have been described from Cambridge, England alone (Smith, 1963a). It is not possible to say as yet whether the many cytoplasmic polyhedroses recorded are all due to different viruses. This is highly unlikely in view of the cross transmissibility of many of these viruses (see Chapter VII).

The only insect order other than the Lepidoptera in which the occurrence of cytoplasmic polyhedroses has so far been recorded is the Neuroptera. The larvae of certain species of lacewing flies, *Hemerobius* and *Chrysopa* spp., are susceptible apparently both to their own virus and viruses from other insect species (Sidor, 1960; Smith, 1963b).

41

A. The Polyhedra

a. SIZE AND STRUCTURE

The cytoplasmic polyhedra in the different diseases vary greatly in size and shape; in some cases they tend to be very large as in the larvae of *Bombyx mori* L., *Ourapteryx sambucaria* L., and *Estigmene acrea;* when large the polyhedra in the latter insect sometimes lose their many-sided character and appear almost spherical (Fig. 9).

In some species such as *Antheraea pernyi* Guérin-Ménéville, for example, polyhedra may be both very large and extremely small, the small polyhedra being at the limit of resolution of the optical microscope. As in the case of the nuclear polyhedra, there may be considerable variation in size of the polyhedra in adjacent cells but each cell seems to have polyhedra of uniform size (Tanada, 1960).

Aruga and Israngkul (1961) have carried out experiments on the factors controlling the size of the cytoplasmic polyhedra in the silkworm. They consider that the size of the polyhedra depends upon the time interval between inoculation and appearance of symptoms; in other words, the longer the time elapsed the larger are the polyhedra formed.

Further work on these lines (Aruga *et al.*, 1963) suggests that the size may vary according to the exact location of the polyhedra in the midgut. Thus, when third-, fourth-, and fifth-instar larvae of *B. mori* were infected *per os* with cytoplasmic polyhedrosis virus, smaller polyhedra were produced in the cells of the caudal portion of the midgut and larger polyhedra in the cephalic portion. However, this did not hold true for first- or second-stage larvae similarly infected.

In cases where a latent infection was stimulated into multiplication by heat the same difference in size of the polyhedra in the caudal and cephalic portions of the midgut was observed, but this was not the case if the latent virus was stimulated into action by low temperatures. The results after chemical induction were inconclusive.

The multiplication of a cytoplasmic polyhedrosis virus in silkworms under conditions of starvation has been investigated. The number of polyhedra formed in the starved larvae was generally slightly less than that in normally fed larvae and the mean size of the polyhedra was correspondingly smaller (Aizawa and Furuta, 1962).

The development of the polyhedra in a cytoplasmic polyhedrosis of *Thaumetopoea pityocampa* Schiffermüller has been described by Xeros

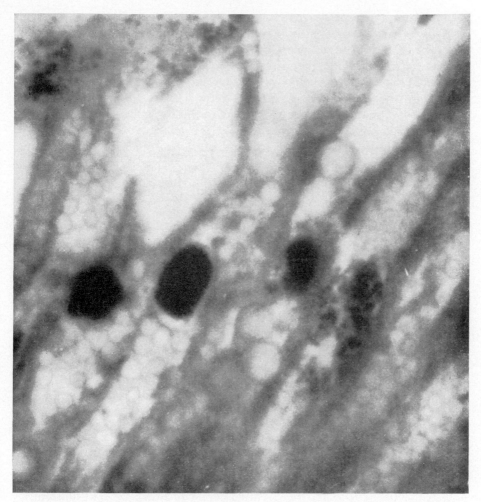

FIG. 9. Photomicrograph of a section through midgut cells of a larva of *Estigmene acrea* affected with a cytoplasmic polyhedrosis; note the large, rounded polyhedra and the wide range of size. Magnification, × 1500.

(1956). Before the formation of the polyhedra, bodies which are apparently virogenic stromata appear in the epithelial cells of the gut. Later, polyhedra of varying size, but mostly rather small, form sparsely over the virogenic stromata. Further development proceeds with the continuing growth of the virogenic masses and with the increase in the number and size of the polyhedra at their surfaces. During this stage of the disease, large pores develop within the virogenic masses, and polyhedra also arise and develop within these pores. In the final stages of the disease the polyhedra reach a size of 1.5 μ in diameter.

Not very much work has been done on the ultrastructure of cytoplasmic polyhedra but Bergold and Suter (1959) have made electron microscope studies of thin sections of osmic acid-fixed material from *Bombyx mori* L., *Arctia villica* L., and *Dasychira pudibunda* L. These showed a molecular lattice in a simple cubic packing about 74 Å, 53 Å, and 57 Å respectively, from center to center. The lattice is not disturbed by the presence of the near-spherical virus particles.

Osmic acid tends to dissolve the virus particles inside the polyhedra if it reaches them, and may give the erroneous impression that the virus particles are composite.

It has been possible to reveal the paracrystalline lattice in cytoplasmic polyhedra by other means than by thin sections. Figure 10 shows part of the crystalline protein of a cytoplasmic polyhedron from *Antheraea pernyi* after treatment at an alkaline pH (Smith and Hills, 1962).

b. PHYSICOCHEMICAL PROPERTIES

Not much is known about the physicochemical properties of the cytoplasmic polyhedra. They are insoluble in water but seem to deteriorate in it and become etched. Their reaction to weak alkali differs somewhat from that of the nuclear polyhedra. They dissolve less easily than the latter and tend to leave behind a matrix full of round holes in which the virus was occluded. There is no "membrane" such as is left behind when the nuclear polyhedra are dissolved in alkali (Figs. 11 and 12).

According to Wittig *et al.* (1960) the cytoplasmic polyhedra from the alfalfa caterpillar are soluble in boric acid and carbonate buffers at pH 10.2. Polyhedra held for several weeks tend to break up into uniformly large particles, and then into smaller particles of different sizes. The reactions of the polyhedra to alkali is further dealt with in the isolation of the virus.

c. CHEMICAL COMPOSITION

As already mentioned in dealing with the nuclear polyhedra, the amino acid composition of some polyhedron proteins and viruses of the silkworm *B. mori* L. have been determined by means of the Beckman-

FIG. 10. Electron micrograph of part of the crystalline protein of a cytoplasmic polyhedron from a larva of *Antheraea pernyi*. Note three virus cores and a single outer coat. Magnification, × 120,000.

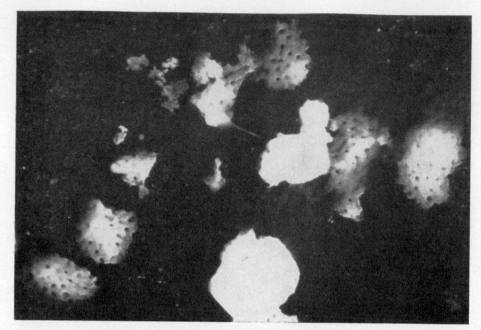

Fɪɢ. 11. Electron micrograph of cytoplasmic polyhedra from a larva of *Phlogophora meticulosa* after treatment with weak sodium carbonate. Note the partially dissolved polyhedra, the empty sockets which had contained the near-spherical virus particles, and the lack of a membrane. Compare with Fig. 2. Magnification, × 28,000.

Spinco Model 120 amino acid analyzer (Kawase, 1964) (see Tables I and II, pp. 26–27). In addition to the nuclear polyhedra, two types of cytoplasmic polyhedra, hexahedra and icosahedra, were examined. The latter type contained much more leucine and isoleucine but less tyrosine and phenylalanine than the former. Both types of polyhedra had very similar patterns of amino acid composition even though their shapes were markedly different.

It has been pointed out that the cytoplasmic polyhedra vary in size according to their position in the midgut (Aruga *et al.*, 1963). The amino acid composition of the polyhedra formed in the caudal portion of the midgut shows much more similarity in pattern to that of the polyhedra formed in starved larvae than to the amino acid composition of polyhedra formed in the cephalic portion of the midgut.

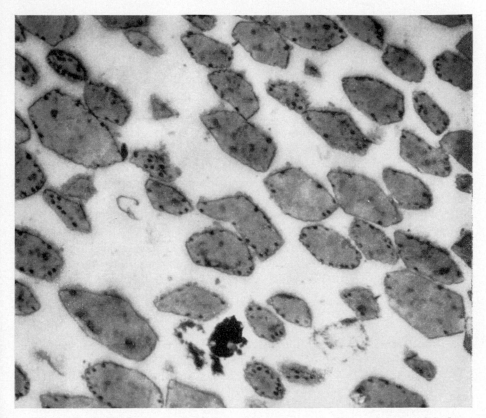

FIG. 12. Electron micrograph of thin sections of very small cytoplasmic polyhedra from a larva of *Hyloicus pinastri*. Note the virus particles, stained black with osmic acid, lying in sockets on the surface. Magnification, × 26,000.

B. Symptomatology and Pathology

The outward appearances of lepidopterous larvae affected respectively with nuclear and cytoplasmic polyhedroses differ considerably, and this is mainly due to the difference in the organs attacked and the sites of virus multiplication. As we have already seen in the nuclear polyhedroses, the skin is one of the main organs attacked, and this is responsible for the characteristic limp cadavers which hang down from the food plant, to which they remain attached by their abdominal legs.

Larvae infected with cytoplasmic polyhedroses lag greatly behind normal larvae in their development. There is no outstanding symptom, at first, but infected individuals can be recognized by their small size, loss of appetite, and sometimes disproportionately large head or long bristles. In some cases a color change develops in a late stage of the disease; larvae of *Pieris rapae*, the small white butterfly, may appear chalky white, particularly on the ventral side. Larvae of *Phlogophora meticulosa* Linn., the angleshades moth, frequently show white patches through the skin in the region of the midgut. These changes are due to the accumulation of masses of polyhedra which can be observed through the integument.

In the late stages of infection, polyhedra are frequently regurgitated or voided in large quantities with the feces.

A peculiar symptom apparently connected with a cytoplasmic polyhedrosis has been observed in full-grown larvae of *A. pernyi*. It occasionally happens that the hindgut is completely extroverted; this extroversion is possibly caused by a blocking of the alimentary canal accompanied by persistent effort on the part of the larva to void the polyhedra.

On opening a larva which has died of a cytoplasmic polyhedrosis, the abnormal state of the alimentary canal, especially the midgut, is at once apparent. Instead of the translucent, pale-green organ of healthy larvae, the gut is opaque and pale yellow or milky in appearance owing to the huge numbers of polyhedra. These sometimes become liberated as a milky fluid during the dissection procedure.

Histological examination of the gut in the early stages of the disease reveals the formation of the polyhedra in the midgut, mainly in the epithelial cells. As infection develops, however, polyhedra spread through the alimentary canal; in the case of a cytoplasmic polyhedrosis of *Estigmene acrea* (Drury), the salt-marsh caterpillar, longitudinal sections of the entire caterpillar show the polyhedra present in almost every cell of the fore-midgut, and hindgut.

The histological changes which take place in the midgut of the processionary caterpillar, or armyworm, *Thaumetopoea wilkinsoni* Tams, have been studied by Harpaz *et al.* (1965). Sections of diseased caterpillars, stained with methylene blue, revealed marked hypertrophy of midgut epithelial cells which contained numerous stained polyhedra. At first these are small and seem to occur only along the epithelial cell wall bordering on the midgut lumen. Later the cell becomes filled with polyhedra and the nucleus begins to disintegrate; next the cell increases greatly in size, assumes a triangular shape and finally bursts, the polyhedra being discharged into the midgut lumen.

As in the case of *O. sambucaria* the polyhedra are almost spherical and measure 1.5 to 2.5 μ in diameter.

These histological changes were noticed only in the columnar cells of the intestinal epithelium and not in the goblet or regenerative cells, a fact already commented on by Xeros (1956) in a study of a cytoplasmic polyhedrosis in *Thaumetopoea pityocampa*.

C. Isolation of the Virus

One of the difficulties of isolating the virus particles in any quantity from the cytoplasmic polyhedra has been to find a technique for liberating the virus particles from their occluding crystals. Unlike the rod-shaped viruses contained in nuclear polyhedra, the viruses from cytoplasmic polyhedra dissolve very readily in weak alkali. Unless care is taken, all that remains after treatment is the polyhedral matrix, honeycombed with the sockets in which the virus particles have rested (Fig. 11).

A method sometimes recommended for the extraction of the polyhedra is simply to put the infected caterpillars into a flask of water and allow normal decomposition to proceed; in time the polyhedra sediment to the bottom of the flask. It has been shown, however (Hills and Smith, 1959), that this is an unsatisfactory method because the polyhedra themselves deteriorate and become etched, while the virus particles embedded in the surface of the polyhedra are lost (Fig. 13). A better method is to open the larvae and plunge them immediately into distilled water; the alimentary canals are then removed under water and the polyhedra extracted. This method gives undamaged polyhedra with the individual virus particles on the surface of the crystal still covered with a layer of protein (Fig. 14).

The following techniques have proved effective in extracting the virus from several types of cytoplasmic polyhedra. A suspension of polyhedra is dialyzed against a weak sodium carbonate solution, adjusted to pH 10.0, for 48 hours, then against water for 48 hours. The extract, after cleaning at 10,000 rpm for 15 minutes in the centrifuge and reduced in bulk, is subjected to gradient centrifugation through sucrose (Brakke, 1951). A clearly defined band is obtained, and a preparation of pure virus results.

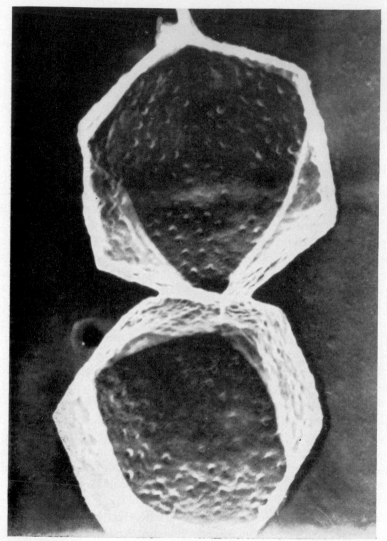

FIG. 13. Carbon replica of two etched cytoplasmic polyhedra from *Antheraea mylitta;* the virus particles have been removed. Note the hexagonal shape of the pits on the crystal surface. Magnification, × 35,000.

An alternative method is to dissolve 50 mg of polyhedra in 1.0 ml of 2% sodium carbonate solution for 30 seconds. The bulk of the liquid is increased to 30 ml by the addition of double-distilled and boiled water and centrifuged at 36,000 rpm for 30 minutes. The pellet is then sus-

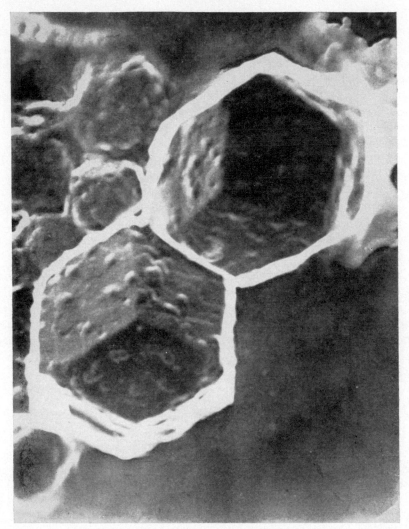

Fig. 14. A carbon replica similar to that in Fig. 13, but with the virus particles still *in situ*. Note the protein cover to the sockets. Magnification, × 35,000.

pended in distilled water and cleaned by centrifugation at 10,000 rpm for 15 minutes. The supernatant is reduced in bulk and subjected to gradient centrifugation through sucrose; this gives a clearly defined band. The band is removed, increased in bulk to 5 ml with water, and centrifuged for 30 minutes at 36,000 rpm; this gives a pellet of pure virus.

This standardized technique has been applied to the cytoplasmic polyhedra of the larvae of the following insects, *Arctia caja* Linn., *Antheraea mylitta* Drury, *A. pernyi* Guérin-Ménéville, *Nymphalis antiopa* Linn., *Vanessa cardiu* Linn., and *Calophasia lunula* (Hufnagel). For the isolation of the virus from these species the optimum conditions appear to be 0.5% sodium carbonate for 1.25 minutes for the polyhedra from A. *caja,* 2% for 30 seconds for A. *mylitta,* 1.0% for 2 minutes for A. *pernyi,* 1.0% for 1.5 minutes for *N. antiopa,* and 1.0% for 1 minute for V. *cardui* and C. *lunula* (Hills and Smith, 1959).

The following technique was used by Hosaka and Aizawa (1964) for isolating a cytoplasmic polyhedrosis virus from the silkworm *B. mori.* Larvae were first fed on polyhedra of the tetragonal type, and the midguts dissected out of larvae showing typical disease symptoms. They were vigorously shaken in distilled water and strained through cheesecloth; then the polyhedra were sedimented by centrifugation at 3000 rpm for 5 minutes.

To isolate the virus, 1 volume of a suspension of polyhedra was mixed with 5 volumes of a mixture of 0.5 M NaCl and 0.05 M Na_2CO_3, and left to stand for 1 hour at room temperature. The suspension was then diluted with distilled water and centrifuged at 3000 rpm for 15 minutes. The clarified supernatant was further centrifuged at 40,000 rpm for 40 minutes; the resulting pellet was resuspended in a small quantity of distilled water.

D. Morphology and Ultrastructure of the Virus Particle

It was first suggested by Kaesberg (1956) that the small plant viruses were not really spherical, but were icosahedral in shape, possessing twenty sides, and it now seems as though icosahedron is the keyword in the contemplation of the "spherical" viruses.

Air-dried preparations of virus particles under the electron microscope are likely to show distortion and to appear smaller than they actually are. Accurate estimation, therefore, of the size and contour can only be made on freeze-dried specimens. Electron micrographs of a particle which is an icosahedron will show a silhouette of five or six sides, and this is more clearly demonstrated in the case of the *Tipula* iridescent virus (TIV) which is dealt with in Chapter V. Some pictures, however, of virus particles from a cytoplasmic polyhedrosis affecting *Antheraea mylitta* show not only the six-sided contour of the particle but also a similarly shaped socket in the polyhedral crystal from which the virus has come. Similar pictures have been taken of a virus from *Automeris io* (Smith, 1958, 1964). The polyhedral shape of the virus sockets in the crystal has been further demonstrated by means of carbon replicas of the polyhedra after removal of the virus particles (Fig. 13).

Another method of demonstrating the contour of the virus particle is by means of shadow analysis. This is described in Chapter V in the section on the *Tipula* iridescent virus; suffice it to say here that the technique has demonstrated that the cytoplasmic polyhedrosis virus of *A. mylitta* is an icosahedron, although the issue is complicated by the very small size of the virus in comparison with the roughness of the substrate on which the shadows are cast (Hills and Smith, 1959).

So far we have been concerned only with the topography of the virus particle; metal shadowing can only be used to this end. To learn more about the details of the structural units it is necessary to use the negative staining technique with phosphotungstic acid (Brenner and Horne, 1959). The over-all picture of a virus particle, derived from various methods of study, is of a thread of nucleic acid—ribonucleic acid (RNA) in plant viruses, deoxyribonucleic acid (DNA) and RNA in the animal viruses and phages—enclosed in a protein coat consisting of large numbers of identical protein subunits; in the small viruses these are the only components.

Crick and Watson (1956) suggested that all small viruses are built up on a framework of identical protein subunits packed together in a regular manner. Klug and Caspar (1960) point out that this means that a virus particle can be constructed out of subunits in only a limited number of ways. Determination of the symmetry gives considerable insight into the substructure of a virus particle.

By means of the negative staining technique it is now possible to count the protein subunits on the face of the particle, and the number of morphological units can be found by adding some combination of the numbers twelve, twenty, thirty, and sixty or a multiple of sixty appropriate for the particular clustering arrangement (Klug and Caspar, 1960).

In studies of the cytoplasmic polyhedrosis of *Antheraea pernyi* the polyhedra have been split up by a minimum pH change into the virus and its protein in solution. These studies have shown that the virus particles consist of an inner core and an outer protein shell. The inner core contains twelve large subunits and the outer shell is made up of a large number of much smaller subunits, but this number is not yet known (Figs. 15, 16). The crystalline lattice of the polyhedral protein is shown in Fig. 10.

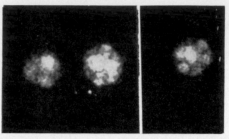

Fig. 15. "Cores" of three virus particles from a cytoplasmic polyhedrosis of *Antheraea pernyi*. Each core is an icosahedron consisting of 12 subunits. Magnification, × 120,000.

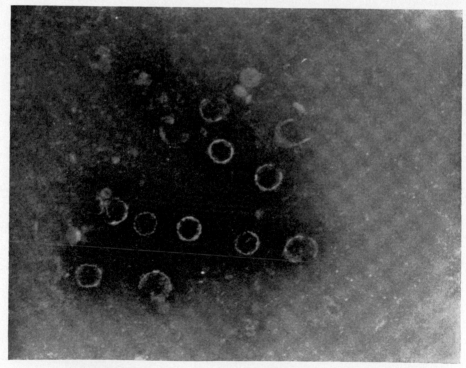

Fig. 16. The outer coats of the virus particles shown in Fig. 15. Magnification, × 100,000.

There is then a considerable body of evidence that the virus particles of the cytoplasmic polyhedroses are icosahedra. The first account of the ultrastructure of the virus is given by Hosaka and Aizawa (1964) who have made an electron microscope study of a cytoplasmic polyhedrosis virus from the silkworm *Bombyx mori*. They find the shape to be icosahedral with two concentric icosahedral shells. Each shell has twelve subunits, localized at twelve vertices of the icosahedron, and twelve tubular structures connect the corresponding subunits of each shell. The subunit of the outer shell is a hollow pentagonal prism from which a projection, consisting of four tubes, protrudes. There is said, also, to be a membranous structure surrounding the subunits (capsomeres). The diameter of the virus varies between 60 and 70 mμ. In the complete virus particle an internal structure approximately 70% of the diameter of the whole particle is present and it is connected to each subunit on the surface shell by a tubular structure. Two structural models of the virus as conceived by Hosaka and Aizawa are reproduced in Fig. A.

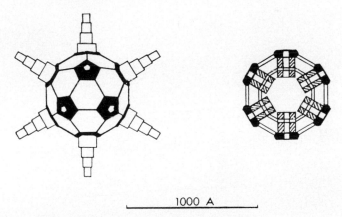

1000 A

Fig. A. Structural models of a cytoplasmic polyhedrosis virus. Left, surface structure; right, internal structure. The front three projections on the surface of the structure are omitted in order to show the structure of pentagonal disks on the vertices. (From Hosaka and Aizawa, 1964.)

In an unpublished study of a cytoplasmic polyhedrosis of the noctuid larva, *Autographa brassicae* Riley, the writer has confirmed the presence of the projections (Fig. 17). Indeed the morphological similarity of this virus to that described above is remarkable. It was originally suggested that the empty shell of the *Tipula* iridescent virus (TIV) consisted of

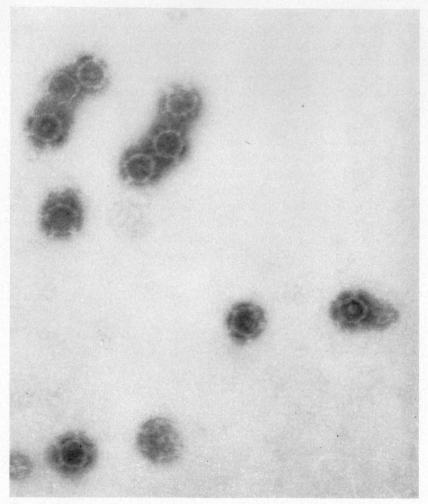

FIG. 17. Virus particles from a cytoplasmic polyhedrosis of a noctuid larvae *Autographa brassicae* Riley. Note the projections. Magnification, × 100,000.

two concentric icosahedra (Hills and Smith, 1959). However, since concentric icosahedra are considered unlikely, it was later suggested that the apparent second membrane in the TIV particle might be the second row of protein subunits.

In preparing the virus for electron microscopy, as shown in Fig. 17, caterpillars of *Autographa brassicae,* in a fairly advanced state of the disease, were induced to void polyhedra directly onto the microscope grid. By this means a small quantity of pure polyhedra was obtained; the grids were then treated with Na_2CO_3 to dissolve the polyhedra, and the particles were washed, and stained with phosphotungstic acid.

E. Chemistry of the Virus

As regards the nucleic acid content of the cytoplasmic polyhedrosis viruses, Xeros (1962) has carried out chromatographic and electrophoretic analysis of such viruses from *Laothoe populi* Linn. This showed that they contained 0.9% RNA and no DNA. Chromatographic analysis of similar viruses from five further species showed the presence in them of similar quantities of RNA and the absence of DNA.

Kawase (1964) has studied the amino acid composition of several polyhedrosis viruses and their polyhedra, the values for each amino acid being expressed as percentages of the total amino acid recovered. The three viruses studied, two types of cytoplasmic polyhedroses and one nuclear polyhedrosis all affecting the silkworm, showed considerable similarity in the pattern of their amino acid composition just as in the case of the polyhedron proteins. However, there was a little variation in some amino acid values, e.g., histidine, arginine, threonine, glutamic acid, proline, alanine, and tyrosine.

REFERENCES

Aizawa, K., and Furuta, Y. (1962). Multiplication of nuclear and cytoplasmic polyhedrosis viruses in starved larvae of the silkworm *Bombyx mori* (L). *J. Insect Pathol.* 4, 465–468.

Aruga, H., and Israngkul, A. (1961). Studies on the size of cytoplasmic polyhedra of the silkworm, *Bombyx mori* L. *Nippon Sanshigaku Zasshi* 30, 119–125.

Aruga, H., Yoshitake, N., and Watanabe, H. (1963). Some factors controlling the size of the cytoplasmic polyhedron of *Bombyx mori* (Linn.). *J. Insect Pathol.* 5, 72–77.

Bergold, G. H., and Suter, J. (1959). On the structure of cytoplasmic polyhedra from some lepidoptera. *J. Insect Pathol.* 1, 1–14.

Brakke, M. K. (1951). Density gradient-centrifugation: a new separation technique. *J. Am. Chem. Soc.* 73, 1847.

Brenner, S., and Horne, R. W. (1959). A negative staining method for high-resolution electron-microscopy of viruses. *Biochim. Biophys. Acta* 34, 103–110.

Crick, F. H. C., and Watson, J. D. (1956). Structure of small viruses. *Nature* 177, 473–474.

Harpaz, I., Zlotkin, E., and Ben Shaked, Y. (1965). On the pathology of cytoplasmic and nuclear polyhedroses of the Cyprus processionary caterpillar, *Thaumetopoea wilkinsoni* Tams. *J. Invert. Pathol.* 7, 15–21.

Hills, G. J., and Smith, K. M. (1959). Further studies on the isolation and crystallization of insect-cytoplasmic viruses. *J. Insect Pathol.* 1, 121–128.

Hosaka, Y., and Aizawa, K. (1964). The fine structure of the cytoplasmic polyhedrosis virus of the silkworm *Bombyx mori* Linn. *J. Insect Pathol.* 6, 53–77.

Ishimori, J. (1934). Contribution à l'étude de la grasserie du ver à soie (*Bombyx mori*). *Compt. Rend. Soc. Biol.* 116, 1169.

Kaesberg, P. (1956). Structure of small "spherical" viruses. *Science* 124, 626.

Kawase, S. (1964). The amino-acid composition of viruses and their polyhedron proteins of the polyhedroses of the silkworm, *B. mori* L. *J. Insect Pathol.* 6, 156–163.

Klug, A., and Caspar, D. L. D. (1960). The structure of small viruses. *Advan. Virus Res.* 7, 225–325.

Lotmar, R. (1941). Die Polyederkrankheit der Kleidermotte (*Tineola bisselliella*). *Mitt. Schweiz. Entomol. Ges.* 18, 372–373.

Sidor, C. (1960). A polyhedral disease of *Chrysopa perla* L. *Virology* 10, 551.

Smith, K. M. (1958). The morphology and crystallization of insect cytoplasmic viruses. *Virology* 5, 168–176.

Smith, K. M. (1963a). The cytoplasmic virus diseases. *In* "Insect Pathology" (E. A. Steinhaus, ed.), Vol. 1, p. 478. Academic Press, New York.

Smith, K. M. (1963b). The arthropod viruses. *In* "Viruses, Nucleic Acids, and Cancer," p. 77. Williams and Wilkins, Baltimore.

Smith, K. M. (1964). Virus diseases of arthropods. *In* "Plant Virology" (M. K. Corbett and H. D. Sisler, eds.), pp. 439–456. Univ. of Florida Press, Gainesville, Florida.

Smith, K. M., and Hills, G. J. (1962). Ultrastructure and replication of insect viruses. *Proc. 5th Intern. Conf. Electron Microscopy, Philadelphia, 1962* Vol. 2, Art. V-1 Academic Press, New York.

Smith, K. M., and Wyckoff, R. W. G. (1950). Structure within polyhedra associated with virus diseases. *Nature* 166, 861.

Smith, K. M., and Xeros, N. (1953). Studies on the cross-transmission of polyhedral viruses: Experiments with a new virus from *Pyrameis cardui*, the painted lady butterfly. *Parasitology* 43, 178–185.

Smith, K. M., Wyckoff, R. W. G., and Xeros, N. (1953). Polyhedral virus diseases affecting the larvae of the privet hawk moth. *Sphinx ligustri* (Linn.) *Parasitology* 42, 287–289.

Tanada, Y. (1960). A nuclear polyhedrosis virus of the lawn armyworm, *Spodoptera mauritia* (Boisd.) (Noctuidae). *Proc. Hawaiian Entomol. Soc.* 17, 304–308.

Wittig, G., Steinhaus, E. A., and Dineen, J. P. (1960). Further studies of the cytoplasmic polyhedrosis virus of the alfalfa caterpillar. *J. Insect Pathol.* 2, 334–345.

Xeros, N. (1952). Cytoplasmic polyhedral virus diseases. *Nature* 170, 1073.

Xeros, N. (1956). The virogenic stroma in nuclear and cytoplasmic polyhedroses. *Nature* 178, 412–413.

Xeros, N. (1962). The nucleic acid content of the homologous nuclear and cytoplasmic polyhedroses virus. *Bichim. Biophys. Acta* 55, 176–181.

CHAPTER IV

The Granuloses

Introduction

The disease known as *granulosis* was first detected by a Frenchman, Paillot (1926), in the larvae of the large white butterfly, *Pieris brassicae* Linn. This, of course, was before the days of the electron microscope, but Paillot recognized that the disease was a new one with a different type of causative agent from that of the polyhedroses and called it *pseudo-grasserie*. Later he discovered a similar disease in a "cutworm," the larva of a noctuid moth *Agrotis segetum* Schiffermuller (Paillot, 1934). He also described two more diseases in the same host (Paillot, 1935, 1936, 1937), and because he thought he could discern differences, he gave them the names pseudograsserie 1, 2, and 3. It is highly probable, however, that all three diseases were caused by the same virus.

So far as is known at present the granulosis disease is found only in larvae, and occasionally pupae, of the Lepidoptera. It is common in the larvae of *P. brassicae* and *P. rapae,* and even more so in the "cutworms," the larvae of noctuid moths. Infected cutworms may frequently be found lying in pathways and similar places, having apparently crawled into the open to die.

The disease was rediscovered by Steinhaus (1947) in the variegated cutworm *Peridroma margaritosa* Haworth, and in 1948 Bergold described a similar disease in the pine shoot roller, *Choristoneura murinana* Hubner, and demonstrated the virus nature of the causative agent with the electron microscope.

In order to avoid confusion it will be well to give at the outset a definition of the various envelopes etc., which surround the virus particle. There is first of all the actual virus material, the nucleoprotein, which

59

is in the shape of a slightly curved rod enclosed in the *intimate membrane*. This in turn is enclosed in an *outer membrane*, sometimes described erroneously as a "developmental" membrane. All this is occluded inside a protein crystal which, itself, is sometimes enclosed in a membrane (see Fig. 18).

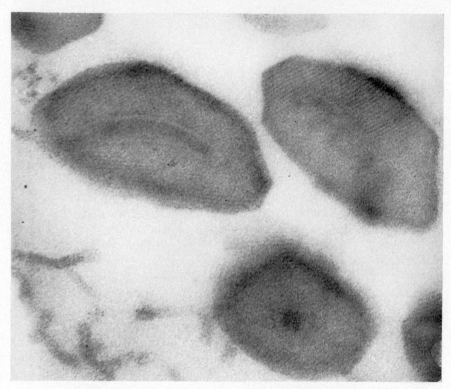

FIG. 18. Section through a capsule from a granulosis disease of *Pieris rapae* L. Note the curved virus rod and its outer membrane within the capsule. The crystalline lattice of the capsule can be seen. Magnification, × 110,000.

There has been some controversy over a suitable name for the protein crystals which are such a characteristic feature of the disease. Bergold (1948) used the expression "Viruskapseln" for them, and "Kapselvirus-Krankheit" for the disease. However, the word "granulosis" has now been generally accepted as being most descriptive of the disease. The choice for the nomenclature of the crystal seems to lie between "granule" or "capsule," neither of which is particularly suitable for a crystal. However, in view of some recent work on this subject to be described in Chapter VI and because there seems to be a consensus of opinion in favor of the word capsule, this is used throughout to designate the crystal container of the virus particle.

A. The Capsules (Granules)

a. SHAPE, SIZE, AND STRUCTURE

The shape of the capsules has been described variously as oval, ellipsoidal, ovoid, or egg-shaped in outline. Lower (1954), in a study of the granulosis affecting *Persectania ewingii* Wester., reported the capsules to be "subovoid, elongate bodies with more or less parallel sides." Their shape should be designated ovocylindrical instead of only uniformly ellipsoidal or ovoid. A certain variation in shape within an individual species seems to be a common feature (Huger, 1963).

Granuloses with aberrant-shaped capsules do occur. Thus, a strain of virus producing only cubic capsules has been described as affecting the spruce budworm *Choristoneura fumiferana* (Clemens). In two serial passages of the virus only the cubic capsules were produced (Stairs, 1964). Figure 18 gives a good impression of the general appearance of the capsules.

Measurements of the actual size of the capsules from one or two different granuloses are available. In the armyworm *Pseudaletia unipuncta* Haworth, 50 capsules selected at random ranged in width from 201.0 to 307.7 mμ with a mean of 253.7 ± 2.34 mμ. Their lengths varied from 439.6 to 602.9 mμ with a mean of 511.2 ± 6.03 mμ (Tanada, 1959). In the larva of the codling moth, *Carpocapsa pomonella* L., the length was 393.9 mμ ± 4.29 mμ; width was 207.7 mμ ± 9.76 mμ (Tanada, 1964).

Wasser and Steinhaus (1951) give the average size of the capsules in the red-banded leaf roller, *Argyrotaenia velutinana* (Walker), as approximately 160 mμ wide by 315 mμ long.

In the cubic capsules mentioned above (Stairs, 1964) the average measurements are 150 × 350 mμ; there seems to be a wide variation in size, the average being about 1 μ but many are less than 0.5 μ and others more than 2 μ.

A general average of the sizes of capsules known to date is given by Huger (1963) as ranging from 300 to 511 mμ in length and 119 to 350 mμ in width.

Ultrathin sections of capsules when examined under the electron microscope readily reveal the macromolecular paracrystalline lattice (Fig. 18). According to Bergold (1959) the molecules are in a cubic arrangement. In capsules of *C. murinana* they have dimensions of $57 \times 57 \times 229$ Å; the sides of the unit cell measure 57 Å, and the angle (Y) is $120°$. As in the nuclear polyhedra there seems to be no disturbance in the molecular lattice from the presence of the virus rod.

b. PHYSICOCHEMICAL PROPERTIES

The capsules are completely insoluble in water or alcohol and are apparently resistant to enzymic action during the decomposition of the host larvae. They are, however, readily soluble in weak alkalis or strong acids, and, as we shall see later, the former are used in the liberation of the occluded virus.

According to Bergold (1963) the capsules of *C. murinana* have a density of 1.279. When dissolved in weak alkali two components are formed; the main molecule has a sedimentation constant, S_{20}, of 11.8 and a molecular weight of 300,000, while the split component has a sedimentation constant, S_{20}, of 3.45 and a molecular weight of about 60,000.

c. CHEMICAL COMPOSITION

There is not much information on the chemical composition of the capsules. Wellington (1951, 1954) has studied the amino acids of the virus and the inclusion-body protein from a granulosis disease of *C. murinana* (Hb.). She found the following amino acids in the acid hydrolyzates of both virus and capsular protein: cysteic acid, aspartic acid, glutamic acid, serine, theonine, alanine, tyrosine, methionine, histidine, lysine, arginine, proline, valine, leucine, isoleucine, phenylalanine, and glycine. Tryptophan was determined in the unhydrolyzed samples.

B. Symptomatology and Pathology

The external symptoms of granulosis in lepidopterous larvae are somewhat variable depending on whether or not the skin is affected. With one exception known at the present time the fat body is always the main organ attacked. The exception is the larva of *Harrisina brillans* B. and McD. in which the principal site of infection is the midgut epithelium (O. J. Smith *et al.*, 1956).

There is frequently a color change in advanced stages of the disease, the larvae of *Pieris rapae* L., for example, may become white on the ventral surface. Another color change has been noted in a granulosis of the codling moth *Carpocapsa pomonella* L. in which the dorsal integument of the infected larvae is more intensely pink than that of healthy larvae. The ventral integument becomes more opaque as infection progresses (Tanad, 1964). In the buckeye caterpillar *Junonia coenia* Hübner there is a brownish discoloration (Steinhaus and Thompson, 1949) and a yellow color develops in affected larvae of *Harrisina brillans* (O. J. Smith *et al.*, 1956). A condition strongly reminiscent of the nuclear polyhedroses develops in affected larvae of *Pieris brassicae* L., the large white butterfly. In this species the epidermis is usually attacked, the body contents become liquefied and the larvae hang head downwards or in an inverted V (Fig. 19). Occasionally, however, infected larvae of this species occur in which the virus is confined to the fat body (K. M. Smith and Rivers, 1956).

In the case of a granulosis of the cabbage looper, *Trichoplusia ni* (Hübner), infection is restricted to the fat body with the possible exception of the hemocytes. The hypodermis is apparently unaffected (Hamm and Paschke, 1963).

In cutworms (Noctuidae) there is generally a whitish appearance of the integument, sometimes on the ventral side, but as the disease progresses the whole larva takes on a whitish cast. At a late stage of the disease, or after death, a chalky white liquid may ooze through the skin. This fluid, of the consistency of thin cream, contains huge numbers of capsules and may dry to a white deposit on the integument.

The external symptoms of granulosis in the larvae of *Harrisina brillans* B. and McD. are somewhat different, as would be expected, since the site of virus multiplication is the midgut epithelium and not the more usual fat body. In infected colonies eggs may fail to hatch and larvae fail to feed. Other symptoms are abnormal feeding, abnormal growth, and coloration of larvae. The abnormal feeding consists of "spotty" feeding; tiny patches of parenchyma are eaten, giving a "peppered" appearance to the leaf. Feeding ceases and the larvae become flaccid and shrunken to about half of normal size, changing in color from lemon yellow through shades of grayish yellow to brown or black. Diarrhea frequently occurs, an unusual symptom for granuloses, and infected larvae tend to wander, trailing behind them a brownish discharge which, on drying, sometimes fastens the larva to the leaf. Insects may also die in

Fig. 19. Three larvae of *Pieris brassicae L.* killed by a granulosis. Note the characteristic appearance of the dead larvae which is often in the form of an inverted V. Compare with Fig. 4. Magnification, × 3.

either prepupal or pupal stages; failure to escape from the cocoon is associated with infected colonies (O. J. Smith *et al.,* 1956).

As regards the histopathology, there was for a time disagreement as to whether the granulosis virus multiplied in the cell nucleus or the cytoplasm. Thus, according to Bird (1958, 1959) the capsules of *C. fumiferana, Eucosma griseana,* and *Pieris rapae* develop only in the cytoplasm of cells, the nuclei of which are still intact. Later, however, (Bird, 1963) he states that the formation of granulosis virus rods begins in the cell nucleus and continues in the cytoplasmic area, in material probably of nuclear origin, following an early rupture of the nuclear membrane. In a granulosis of *Natada nararia* (Moore) the virus was found to multiply in the nuclei of the epidermis and fat-body cells (K. M. Smith and Xeros, 1954). Similarly, nuclear development of the virus has been described in the larvae of *Pieris brassicae* L. (K. M. Smith and Rivers, 1956), in the cabbage looper, *Trichoplusia ni* (Hubner), and in the armyworm *Pseudaletia unipuncta* Haworth (Tanada, 1959).

Huger (1961) used a selective stain for capsules (see Chapter X) which convinced him that in the granulosis of *C. murinana* the capsules develop in the nucleus as well as in the cytoplasm. The changes in the infected cells as observed in the optical microscope have been well summarized by Huger (1963). The first signs of infection develop in the fat body where the cells begin to proliferate leading to more voluminous fat-body lobes, referred to by Paillot (1934, 1935, 1936) as "proliferation cellulaires." (See also Martignoni, 1957; Wittig, 1959). At the same time the nuclei begin to increase in size and the nucleoli disappear.

The nucleus continues to grow until it almost fills the cell much in the same manner as in the nuclear polyhedroses (see Figs. 6, 8). The capsules then become visible in the enlarged nuclei as a cloudy, light-staining mass with a granular appearance; at the same time capsules appear in the cytoplasm. The most characteristic feature, however, is the development of an intensely staining network in the nuclear area as well as in the cytoplasm (Hughes and Thompson, 1951; Wittig, 1959; Huger, 1960). Finally, the nuclear membrane breaks down and the constitutents of nucleus and cytoplasm intermingle.

Huger (1960) suggested that the network might serve as a virogenic stroma and this was investigated with the electron microscope by Huger and Krieg (1960, 1961). The subject is dealt with further in Chapter VI in discussing the replication of viruses.

In a study of a granulosis affecting the cabbage looper, *Trichoplusia ni* (Hübner), Hamm and Paschke (1963) describe a Feulgen-positive network which they consider a virogenic stroma similar to that reported by Huger (1960) and Huger and Krieg (1961). The stroma in *T. ni* eventually spreads over the entire cell, further evidence that the virus multiplies in both the nucleus and the cytoplasm of the infected cell.

A number of oval or nearly spherical vesicles containing both capsules and virus rods seem to be always present in infected cells. These vesicles may be very small, containing two or three capsules, or quite large with

many capsules. They were first observed by Paillot (1937) who called
them "boules hyaline," and have since been seen by a number of workers
(Bergold, 1958; Hughes and Thompson, 1951; Tanada, 1953; Martignoni,
1957). Their significance is not known. An ultrathin section through one
of these vesicles is shown in Fig. 20.

C. Isolation of the Virus

The isolation of the granulosis viruses is a two-step process, just as
with the viruses of the polyhedroses; that is to say that the inclusion
bodies, or capsules, must first be purified and the virus subsequently
liberated from them. Actually the process is very similar in both types
of disease, weak alkali being used in each case to dissolve the purified
inclusion bodies.

The extraction of the capsules from the diseased larvae is easy
enough, and the quantity obtained from a single diseased cutworm is
astonishingly large. If a number of incisions is made in the skin of the
larva and it is then shaken up in distilled water, about 10 ml of a fluid
with the consistency of thin cream is obtained. This consistency is due
to the vast numbers of capsules liberated from the disintegrating fat
body. Centrifugation of this milky suspension for 30 minutes at 5000
rpm sediments the capsules and gives a supernatant with a greenish
opalescence. The supernatant contains numerous long thin virus rods
which branch in an intricate manner; the significance of these rods is
discussed in Chapter VI on the replication of the viruses.

The following technique, which is essentially the same as that used
for isolating viruses from nuclear polyhedroses, is quoted from Bergold
(1964). Five milligrams of inclusion bodies is used for each milliliter
of a solution of 0.004–0.03 M Na_2CO_3 + 0.05 M NaCl. The inclusion
bodies should dissolve at room temperatures in about 1 to 2 hours. During this time the milky suspension becomes opaque. It is centrifuged
for about 5 minutes at 2000–4000 g to sediment insoluble impurities.
There is practically no brownish pellet if the inclusion bodies were pure.
A white sediment indicates that not enough alkali was used. The supernatant is bluish-white with the virus particles suspended in the solution
of inclusion-body protein. This supernatant is now centrifuged for 1 hour
at about 10,000 g. The virus particles collect in a bluish-white pellet. The
supernatant, consisting of a solution of the capsule protein, is discarded
and the virus pellet is suspended in the same volume of CO_2-free dis-

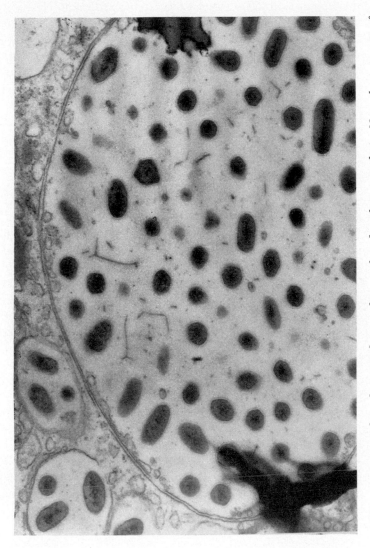

FIG. 20. Section through a vesicle in a larva infected with a granulosis. Note the presence of branching virus rods as well as the capsules. Magnification, × 13,000. (From Smith and Brown, 1965.)

tilled water. The virus particles are sedimented again by contrifuging at 10,000 g for another hour. The supernatant is discarded and the virus suspended in one-seventh of the original volume in CO_2-free distilled water; this gives a bluish-white suspension of pure virus.

If only sufficient virus is required for observation in the electron microscope, the capsules can be dissolved directly on the grid. A drop of capsule suspension is placed on the grid, allowed to stand for a moment and then dried off with filter paper. A solution of alkali is then placed on the grid and drawn off after a period which must be determined by trial and error.

D. Morphology and Ultrastructure of the Virus

There are apparently two different forms of the granulosis viruses and what is thought to be the relationship between them is discussed in Chapter VI. For the moment we are concerned only with their morphology and structure. As we have mentioned previously the capsule contains a short rod, rather thick and slightly curved, which is itself enclosed in an outer membrane. This membrane is separated by a short distance from the intimate membrane which contains the actual virus material. The dimensions of these short rods are available for several granuloses. Tanada (1959) gives the following measurements of 25 naked virus rods from the armyworm *Pseudaletia unipuncta* Haworth: width from 57.8 to 73.5 mμ and a mean of 62.0 mμ with a standard error of ±0.699 mμ, lengths ranged from 367.5 to 441.0 mμ with a mean of 411.6 ± 3.64 mμ. The same worker (Tanada, 1964) gives the average size and standard error of the mean of ten virus particles from the larva of the codling moth, *Carpocapsa pomonella* L., as follows: length 313.5 ± 8.02 mμ, width 50.7 ± 0.30 mμ.

For the aberrant virus strain producing only cubic inclusion bodies in the spruce budworm, *Choristoneura fumiferana* (Clemens), the measurements given by Stairs (1964) are 40×270 mμ. Huger (1963) summarizes the average measurements for granulosis viruses as a whole as having a width ranging from 36 to 80 mμ, and a length from 245 to 411 mμ.

Not very much is known about the ultrastructure of the granulosis virus particles; K. M. Smith and Hills (1962) by means of negative staining detected a closely packed helical structure of the intimate membrane of isolated granulosis virus rods which could not be resolved

FIG. 21. High-resolution electron micrograph of part of a naked granulosis virus of *Pieris brassicae* L., showing a helical structure on the intimate membrane. Magnification, × 400,000.

69

on empty membranes (Fig. 21). A little more is known about the ultra-structure of the long branching threads which seem to be a stage in the replication of the virus (see Chapter VI). These long threads are composite in nature and appear to consist of an outer coat of protein wound round an internal core which is helical in structure (Figs. 22 and 23).

During negative staining with phosphotungstic acid (PTA), the stain may remove part of the protein covering and reveal the inner core. Alternatively, the PTA sometimes enters the rod at a break in the outer coat and delineates the apparent helix inside (Fig. 23).

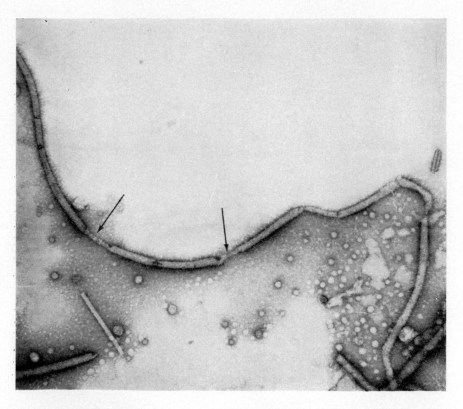

FIG. 22. A long virus rod from a granulosis of a noctuid larva showing the apparently helical core (arrows). Magnification, × 45,000. (From Smith and Brown, 1965.)

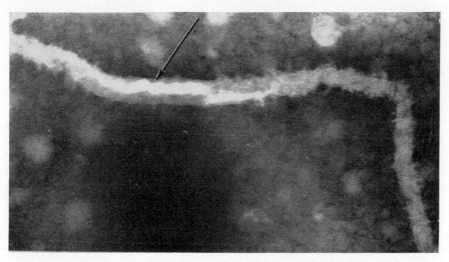

FIG. 23. A long virus rod from a granulosis of a noctuid larva negatively stained with phosphotungstic acid. Note that the stain has penetrated the outer coat and delineated the apparent helix within (arrow). Magnification, × 88,000. (From Smith and Brown, 1965.)

E. Chemistry of the Virus

What little work has been done on the chemistry of the granulosis viruses has been carried out as part of a study of the chemistry of the inclusion-disease viruses as a whole.

Wellington (1951, 1954) has made both qualitative and quantitative studies of the amino acids of a number of insect viruses including the granulosis virus of *Cacoecia murinana* (Hb.), and, as previously mentioned, found the following amino acids in the acid hydrolyzates of a nuclear polyhedrosis virus from *Bombyx mori*, a granulosis virus of *Cacoecia murinana*, and their corresponding polyhedral and capsular proteins: cysteic acid, aspartic acid, glutamic acid, serine, threonine, alanine, tyrosine, methionine, histidine, lysine, arginine, proline, valine, leucine, isoleucine, phenylalanine, and glycine.

Wyatt (1952a,b) has studied the DNA content of granulosis viruses in *Choristoneura murinana* and *C. fumiferana*. He found that they contain the purines adenine and guanine, and the pyrimidines cytosine and thymine, but no methylcytosine or uracil. Although the two viruses come from closely related hosts they differ in their nucleic acid composition.

REFERENCES

Bergold, G. H. (1948). Über die Kapselvirus-Krankheit. Z. *Naturforsch.* 3b, 338–342.

Bergold, G. H. (1958). Viruses of insects. *In* "Handbuch der Virusforschung" (C. F. Hallauer and K. F. Meyer, eds.), Vol. 4, Suppl. 3, pp. 60–142. Springer, Vienna.

Bergold, G. H. (1959). Structure and chemistry of insect viruses. *Proc. 4th Intern. Congr. Biochem., Vienna, 1958* Vol. 7, pp. 95–98. Pergamon Press, Oxford.

Bergold, G. H. (1963). The molecular structure of some insect virus inclusion bodies. *J. Ultrastruct. Res.* 8, 360–378.

Bergold, G. H. (1964). Insect Viruses. *In* Techniques in Experimental Virology (R. J. C. Harris, ed.) Academic Press, New York.

Bird, F. T. (1958). Histopathology of granulosis viruses in insects. *Can. J. Microbiol.* 4, 267–272.

Bird, F. T. (1959). Polyhedrosis and granulosis viruses causing single and double infections in the spruce budworm, *Choristoneura fumiferana* Clemens. *J. Insect Pathol.* 1, 406–430.

Bird, F. T. (1963). On the development of granulosis viruses. *J. Insect Pathol.* 5, 368–376.

Hamm, J. J., and Paschke, J. D. (1963). On the pathology of a granulosis of the cabbage looper, *Trichoplusia ni* (Hübner). *J. Insect Pathol.* 5, 187–197.

Huger, A. (1960). Über die Natur des Fadenwerkes bei der Granulose von *Choristoneura murinana* (Hbn.) (Lepidoptera, Tortricidae). *Naturwissenschaften* 47, 358–359.

Huger, A. (1961). Methods for staining capsular virus inclusion bodies typical of granuloses of insects. *J. Insect Pathol.* 3, 338–341.

Huger, A. (1963). Granuloses of insects. *In* "Insect Pathology" (E. A. Steinhaus, ed.), Vol. 1, p. 538. Academic Press, New York.

Huger, A., and Krieg, A. (1960). Elektronenmikroskopische Untersuchungen zur Virogenese von *Bergoldiavirus calypta* Steinhaus. *Naturwissenschaften* 47, 546.

Huger, A., and Krieg, A. (1961). Electron microscope investigations on the virogenesis of the granulosis of *Choristoneura murinana* (Hbn.) *J. Insect Pathol.* 3, 183–196.

Hughes, K. M., and Thompson, C. G. (1951). A granulosis of the omnivorous looper *Sabulodes caberata* Guenée. *J. Infect. Diseases* 89, 173–179.

Lower, H. F. (1954). A granulosis virus attacking the larvae of *Persectania ewingii* West, W. (Lepidoptera: Agrotidae) in South Australia. *Australian J. Biol. Sci.* 7, 161–167.

Martignoni, M. E. (1957). Conributo alla conoscenza di una granulosi di *Eucosma griseana* (Hübner). (Tortricidae, Lepidoptera) quale fattore limitante il pullalamento dell'insetto nella Engadina alta. *Mitt. Schweiz. Zentralanstalt Forstl. Versuchsw.* 32, 371–418.

Paillot, A. (1926). Sur une nouvelle maladie du noyau au grasserie des chenilles de *P. brassicae* et un nouveau groupe de microorganismes parasites. *Compt. Rend.* 182, 180–182.

Paillot, A. (1934). Un nouveau type de maladie à ultravirus chez les insectes. *Compt. Rend.* **198**, 204–205.

Paillot, A. (1935). Nouvel ultravirus parasite d'*Argrotis segetum* provoguant une proliferation des tissus infectés. *Compt. Rend.* **201**, 1062–1064.

Paillot, A. (1936). Contribution a l'étude des maladies à ultravirus des insectes. *Ann. Epiphyties Phytogenet.* **2**, 341–379.

Paillot, A. (1937). Nouveau type de pseudograsserie observé chez les chenilles d'*Euxoa segetum. Compt. Rend.* **205**, 1264–1266.

Smith, K. M., and Brown, R. M. (1965). On the origin of the long virus threads associated with insect granuloses. *Virology* **27**, 512–519.

Smith, K. M., and Hills, G. J. (1962). Multiplication and ultrastructure of insect viruses. *Proc. 11th Intern. Congr. Entomol., Vienna, 1960* Vol. 2, pp. 823–827, Springer-Verlag.

Smith, K. M., and Rivers, C. F. (1956). Some viruses affecting insects of economic importance. *Parasitology* **46**, 235–242.

Smith, K. M., and Xeros, N. (1954). A comparative study of different types of viruses and their capsules in the polyhedroses and granuloses of insects. *Parasitology* **44**, 400–406.

Smith, O. J., Hughes, K. M., Dunn, P. H., and Hall, I. M. (1956). A granulosis virus disease of the western grape leaf skeletonizer and its transmission. *Can. Entomologist* **88**, 507–515.

Stairs, G. R. (1964). Selection of a strain of insect granulosis virus producing only cubic inclusion bodies. *Virology* **24**, 514.

Steinhaus, E. A. (1947). A new disease of the variegated cutworm, *Peridroma margaritosa* (Haw.) *Science* **106**, 323.

Steinhaus, E. A., and Thompson, G. C. (1949). Granulosis disease in the buckeye caterpillar, *Junonia coenia* Hübner. *Science* **110**, 276–278.

Tanada, Y. (1953). Description and characteristics of a granulosis virus of the imported cabbageworm. *Proc. Hawaiian Entomol. Soc.* **15**, 235–260.

Tanada, Y. (1959). Descriptions and characteristics of a nuclear polyhedrosis virus and a granulosis virus of the armyworm *Pseudaletia unipuncta* (Haworth) (Lepidoptera; Noctuidae). *J. Insect Pathol.* **1**, 215–231.

Tanada, Y. (1964). A granulosis virus of the codling moth. *Carpocapsa pomonella* (Linn.) (Olethreutidae: Lepidoptera). *J. Insect Pathol.* **6**, 378–380.

Wasser, H. B., and Steinhaus, E. A. (1951). Isolation of a virus causing granulosis of the red banded leaf-roller. *Virginia J. Sci.* [N.S.] **2**, 91–93.

Wellington, E. F. (1951). Amino acids of two insect viruses. *Biochim. Biophys. Acta* **7**, 238–243.

Wellington, E. F. (1954). The amino acid composition of some insect viruses and their characteristic inclusion-body proteins. *Biochem. J.* **57**, 334–338.

Wittig, G. (1959). Ein Beitrag zur Histopathologie der Kapselvirose von *Choristoneura murinana* (Hbn.) (Lepidoptera Tortricidae). *Proc. 4th Intern. Congr. Crop. Protect., Hamburg, 1957* Vol. 1, pp. 895–898.

Wyatt, G. R. (1952a). Specificity in the composition of nucleic acids. *Exptl. Cell Res.* Suppl. 2, 201–217.

Wyatt, G. R. (1952b). The nucleic acids of some insect viruses. *J. Gen. Physiol.* **36**, 201–205.

CHAPTER V

The Noninclusion and Miscellaneous
Virus Diseases

PART I. THE NONINCLUSION DISEASES

Introduction

As the name implies there are no crystals, granules, or other inclusion bodies associated with this type of disease. The virus occurs freely in the tissues as do the viruses affecting plants and the higher animals. The number of viruses of this type so far recorded from insects is small, less than a dozen at the time of writing; this small number is probably more apparent than real, arising from a lack of knowledge and investigation. This is understandable in view of the fact that the noninclusion viruses are more difficult to study and need an electron microscope for their definite identification.

Each of the more important noninclusion diseases is dealt with separately, and the same descriptive procedure as that used with the inclusion diseases is repeated, so far as it is applicable. An account is, therefore, given of the host range, the symptomatology and pathology, methods for the isolation of the virus, morphology and ultrastructure of the virus particle, and the chemistry of the virus, in cases where that has been studied. The methods of virus replication are not dealt with here but are discussed in Chapter VI.

A. The *Tipula* Iridescent Virus (TIV)

This unusual and interesting virus was first discovered by the Virus Research Unit at Cambridge, England during routine examination of large numbers of the larvae of a fly, *Tipula paludosa* Meig., for the presence of another virus, that of a nuclear polyhedrosis attacking the

74

same insect (see p. 30). Brief descriptions were later given by Xeros (1954) and Smith (1955).

a. HOST RANGE

Although the original host of this virus is the soil-living dipterous larva of the crane fly *Tipula paludosa* Meig., it is not confined to this insect, and experimentally the virus has been transmitted to a wide variety of insects. A systematic investigation of the host range was carried out by Smith *et al.* (1961), and much of the following information is from that work.

The easiest method of transmitting the viruses of polyhedroses is to feed leaf material contaminated with polyhedra to the experimental insects. However, feeding as a method of transmitting TIV is unsatisfactory, and in many cases is quite unsuccessful. Instead, the insects were injected with a suspension of purified virus, together with an addition to the inoculum of 200 units of penicillin and 200 units of streptomycin per milliliter. The larvae of a few species reacted badly to injection, and these were induced to drink a drop of highly purified virus suspension instead.

The larvae of the following insects, in addition to the natural host *Tipula paludosa* Meig., have been experimentally infected with TIV:

> *Diptera: Tipula oleracea* Linn.; *T. livida* Van der Wulp; two unidentified species of *Tipula; Bibio marci* Linn., St. Mark's fly; *Calliphora vomitoria* Linn.; *Mycetophilus* sp.
>
> *Lepidoptera: Porthetria* (= *Lymantria*) *dispar* Linn.; *Bombyx mori* Linn.; *Hepialus lupulinus* Linn.; *H. humuli* Linn.; *Sphinx ligustri* Linn.; *Pieris brassicae* Linn.; *P. rapae* Linn.; *p. napi* Linn.; *Gonepteryx rhamni* Linn.; *Nymphalis io* Linn.; *Vanessa atalanta* Linn.; *Galleria mellonella* Linn.
>
> *Coleoptera: Tenebrio molitor* Linn.; *Agriotes obscurus* Linn.; *Melolontha* sp.

From the preceding list it will be seen that TIV has been transmitted experimentally to the larvae of seven species of Diptera, twelve species of Lepidoptera, and three species of Coleoptera. The question of the cross transmission of insect viruses is further dealt with in Chapter VII.

The choice of experimental insects was governed largely by what insects were available at the time, and there is little doubt that the experimental host range could be considerably widened. It might be added here that all attempts to infect several species of Orthoptera were unsuccessful.

So far as the writer is aware the above is the first record of the transmission of an insect virus between different orders of insects.

b. Symptomatology and Pathology

A larva of *T. paludosa* affected with TIV can be recognized at once by the change in appearance that it induces in the body of the affected host. Whereas the normal appearance of these larvae is a dark tan, the color of the diseased larvae is a somewhat opalescent blue-indigo. Examination by low-power microscopy shows that the color is particularly intense in the lobes of the hypertrophied fat body; as the disease approaches its termination the color is less intense and is more generally dispersed throughout the body. When a larva in an advanced stage of the disease is opened, a thin white fluid is liberated together with a highly concentrated suspension of the virus which can be recognized by its iridescent colors. The thin white fluid consists of the disintegrated fat body.

In the early stages of the disease before the iridescence is well developed, diagnosis is helped by wetting the surface of the larva in a tube held at an angle so that the light shows up the fat body.

Tipulid larvae affected with the nuclear polyhedrosis also appear white and opaque; however, the brilliant iridescence characteristic of TIV infection is lacking. A white fluid is also liberated from these larvae, but a smear stained with Giemsa solution and examined under the microscope will reveal the presence in large numbers of the crescent-shaped "polyhedra" (see page 30).

The initial site of virus multiplication is the cytoplasmic regions of the fat-body cells, for it is here that the virus is found to be most heavily concentrated. The intracellular virus appears to avoid the nucleus as can be seen from Fig. 24. This is in marked contrast to the behavior of the virus of the nuclear polyhedrosis affecting the same insect, where the rod-shaped virus particles are found within the nuclei of the larval blood cells.

Thin sections of fat body from a larva infected with TIV have an unmistakable appearance when examined with the electron microscope. The cell cytoplasm is completely filled with virus particles which, in an advanced stage of the disease, will already be oriented into a crystalline pattern (Fig. 24). It is these microcrystals formed in the living insect which produce the iridescence.

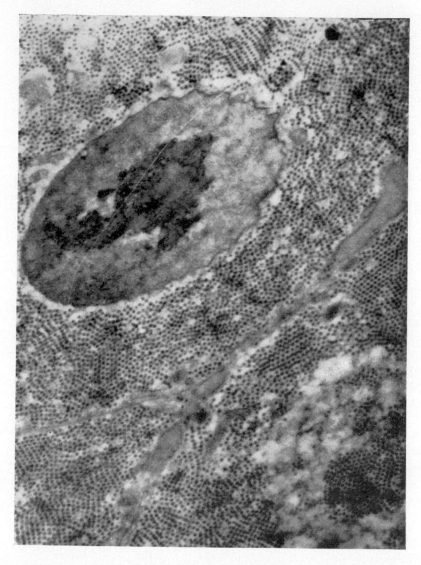

Fig. 24. Electron micrograph of a section through part of a fat-body cell of the larva of *Tipula paludosa* Meig., infected with the iridescent virus. Note that the virus is confined to the cytoplasm and that the particles are oriented. Magnification, × 10,000.

Although the fat body is the initial site of multiplication of the virus, other organs become infected as the disease develops. E. S. Anderson *et al.* (1959) showed the presence of TIV in the skin of *T. paludosa* by means of fluorescent staining, and this finding has been confirmed by electron microscopy. Indeed, the virus seems able to multiply in many types of insect tissue including the muscles, wing buds, legs, and head.

Sometimes groups of virus particles occur together enclosed in a kind of vesicle with double or even treble membranes somewhat reminiscent of the groups of capsules enclosed in a membrane which occur in the granuloses (Fig. 20).

c. ISOLATION OF THE VIRUS

TIV is probably the easiest of all viruses to isolate and purify; this is partly because of the large size of the virus particles and the absence of particles of comparable size in extracts of diseased larval tissue. Another reason for its ease of purification is the astonishingly large quantity of TIV produced by a larva in a late stage of the disease. The dry weights of the extracted, purified virus and of the whole diseased larva have been compared; in the sample measured the virus weighed approximately 25% as much as the entire larva. This yield is easily a record for animal viruses; it is more than twice that reported for tobacco mosaic virus, where the dry weight of the purified virus may be as much as 10% that of the diseased plant leaves. Only the yield of the *T*-even bacteriophage from infected cells of *Escherichia coli* Castellani and Chalmers may be of comparable magnitude.

In order to obtain a purified virus suspension, the infected larvae are first cut up and placed in a beaker of water for several hours. After clarification of the resulting extract by low-speed centrifugation, the virus is obtained essentially pure by the application of two cycles of high- and low-speed centrifugation. Distilled water is a suitable suspending medium.

d. OPTICAL PROPERTIES OF THE VIRUS

The pellets of virus resulting from centrifugation have fascinating optical properties. By transmitted light the pellet appears an orange or amber color; by reflected light it has an iridescent, turquoise appearance. Within the pellet may be seen small regions reflecting the incident light quite brilliantly, giving the entire pellet the appearance of an opal.

When an embedded pellet is sectioned and examined in the electron microscope, the general appearance is that shown in Fig. 25. The pellet

FIG. 25. Electron micrograph of a thin section through a pellet of purified *Tipula* iridescent virus. Note the patterns resulting from sections through small crystalline regions oriented at random. Magnification, × 15,000. (From Williams and Smith, 1957).

is seen to consist of regions of crystallinity, the average diameter of which is 5 to 10 mμ. Since the crystals are randomly oriented with respect to each other, a given section will cut them in different directions with respect to their faces, and the bizarre pattern shown in Fig. 25 will result.

The center-to-center distance of the virus particles in the most closely packed arrays yet found is 1300 Å. The origin of the reflection effects in the pellets is now evident. Bragg reflections result from the periodic particle arrays, and the scale of size of the interparticle spacing is such as to make visible the effects of the Bragg reflections. The change of the reflected color from turquoise to violet upon fixation and dehydration of the pellet is due to shrinkage of the interparticle spacings. The inverse effect can be demonstrated upon rehydrating an unfixed, but partially dried, pellet. At first the pellet is neutral by reflected light, but as it swells upon rehydration the first color to appear is violet, followed by indigo, blue, and finally turquoise (Williams and Smith, 1957).

e. MORPHOLOGY AND ULTRASTRUCTURE OF THE VIRUS PARTICLE

In the early days of the investigation of virus morphology by electron microscopy, but after the advent of metal shadowing, it became apparent that all the virus forms were flattened. This flattening was found to be due to the forces of surface tension. To eliminate this source of morphological artifact, the technique of *freeze-drying* was developed; by the use of this technique it was shown that the "heads" of the *T*-even bacteriophage possessed a distinct polyhedral shape, the general appearance of which was that of a hexagonal prism with pyramidal ends (T. F. Anderson, 1952).

Following upon this it was observed that the contours of some of the so-called spherical particles were not round but six-sided. Kaesberg (1956) attempted to find the polyhedral form of some plant viruses, when freeze-dried, by analysis of shadow shapes. A particular type of polyhedron, if it is regular, should cast a particular type of shadow, if it is cast with a specified orientation with respect to the positions of the vertices of the polyhedral particle. Kaesberg concluded that the shapes of the small viruses investigated by him are best represented by regular icosahedra, figures with twenty sides. One difficulty here is due to the smallness of shadow detail in comparison with the roughness of the film upon which the shadows are cast. This became evident in some attempts to show the icosahedral shape of the virus from the cytoplasmic polyhedrosis of *Antheraea mylitta* by the double shadowing technique (Hills and Smith, 1959).

However, the difficulty brought about by the roughness of the substrate film is obviously reduced if the particle and its associated shadow are large. This is the case with TIV since it is a large virus measuring about 130 mμ in diameter. In spite, however, of the flattening of the TIV particle on drying out of water, it still retains its distinctive noncircular contour. Six-sided contours are the most frequent but five-sided ones also occur (Fig. 26).

In order to prove that the TIV particle is actually an icosahedron,

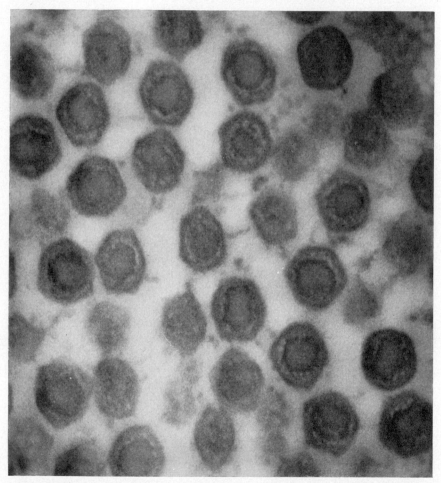

FIG. 26. Electron micrograph of thin sections of the *Tipula* iridescent virus. Note the five- and six-sided contour of the virus. Magnification, $\times$ 60,000.

the double shadowing was carried out as follows. A model of an icosahedron was made and shadowed by two light sources separated 60° in azimuth and oriented so that an apex of the hexagonal contour points directly to each light source. This throws two shadows, one is four-sided and pointed, and the other is five-sided with a blunt end (Fig. 27). A particle of TIV freeze-dried and shadowed in the same way is shown in Fig. 28; the similarity between the shadows thrown is evident. This indicates with fair certainty that the TIV particle is an icosahedron (Williams and Smith, 1958).

The early attempts to investigate the ultrastructure of the TIV particle were directed toward the study of thin sections of the particles under the electron microscope. These sections demonstrated vividly the six-sided appearance of the particles, but they also suggested that each particle was surrounded by a double membrane (Fig. 29). In thin sections of TIV particles from *Tipula paludosa* stained with phosphotungstic acid, and also in whole particles negatively stained, it is possible to make out individual units on the surface of the virus particle. High-resolution micrographs of negatively stained virus from larvae of *Bibio marci* Linn., do show a regular arrangement of the protein subunits, the icosahedral faces being made up of small subunits. Similar subunits have been observed on virus particles obtained from larvae of the white butterfly, *Pieris brassicae* Linn. (Smith and Hills, 1959).

As with a number of other viruses, numerous empty particles of TIV occur, and the significance of these is discussed in Chapter VI, dealing with the replication of insect viruses. Further investigation of these empty particles, using the negative staining technique with phosphotungstic acid (PTA) (Brenner and Horne, 1959) suggested that what appeared to be a second membrane was actually an inner row of protein subunits. Further high-resolution electron microscopy of the empty TIV shells with negative staining shows them to be composed of a large number of protein subunits, tentatively put as 812, with which are found lipids apparently helping to bind the subunits into a rigid structure. The subunits measure 85 Å by 140 Å and are hollow and hexagonal when viewed end-on (Fig. 30). They are arranged to form a 20-sided solid figure (icosahedron), each side being an equilateral triangle (Smith and Hills, 1962).

f. CHEMISTRY OF THE VIRUS

Studies on the chemical composition of TIV have been carried out by Thomas (1961) and the following facts are quoted from his work.

FIG. 27. A model of an icosahedron shadowed by two light sources and oriented so that an apex of the hexagonal contour points directly to each light source. This throws two shadows, one is four-sided and pointed, and the other is five-sided with a blunt end. (From Williams and Smith, 1958.)

The analyses of TIV indicate that it contains 12.4% DNA and 5.2% lipid, most of which is phospholipid. It does not contain any appreciable amount of polysaccharide or RNA; the remainder of the virus, 82.4%, appears to be protein. The DNA of the virus contains only adenine,

FIG. 28. A particle of the *Tipula* iridescent virus frozen-dried and shadowed in the same manner as the model in the preceding figure; the similarity between the shadows thrown is evident. Magnification, × 105,000 (From Williams and Smith, 1958.)

guanine, thymine, and cytosine; there is no methylcytosine or 5-hydroxymethylcytosine. The particle weight of TIV, 1.22×10^9, is the largest so far determined among viruses that are highly uniform in size and shape.

Thomas and Williams (1961) studied the localization of the DNA and the protein in TIV by means of enzymatic digestion and electron microscopy. They found that the DNA is concentrated in the core of

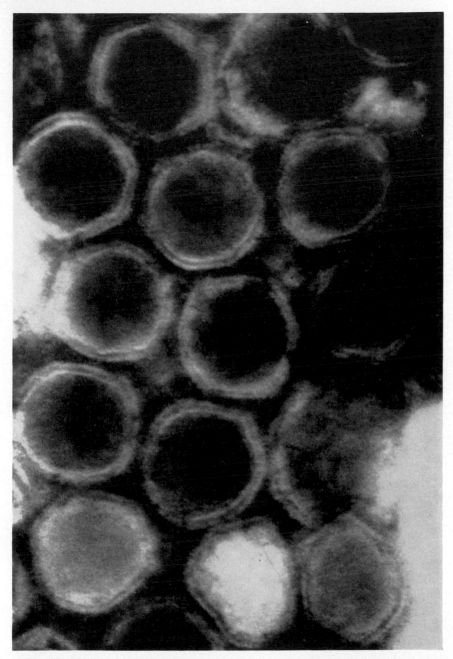

Fig. 29. Electron micrograph of particles of the *Tipula* iridescent virus, negatively stained with phosphotungstic acid. Note the hexagonal shape and apparent second membrane which is now known to be the inner row of protein subunits. Magnification, × 240,000.

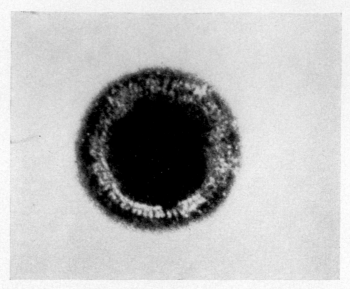

FIG. 30. Single particle of the *Tipula* iridescent virus at very high magnification. Note the protein subunits which are apparently hollow. Magnification, × 300,000.

the particle (90 mμ in diameter) which is estimated to be 30% DNA and 60–65% protein.

B. *Sericesthis* Iridescent Virus (SIV)

A dead larva of the pruinose scarab, *Sericesthis pruinosa* (Dalman) (Coleoptera), was taken in 1962 from pasture land in New South Wales, Australia by Dr. R. J. Roberts. It showed a blue coloration and was sent to Dr. E. A. Steinhaus for examination. It was found to be infected with a virus which was easily transmissible to the larvae of the greater wax moth, *Galleria mellonella* Linn. in which it produced iridescent symptoms similar to those of TIV. A brief note was published by Steinhaus and Leutenegger (1963). Through the courtesy of Dr. Steinhaus a sample of this virus was sent to the writer, who compared it on the electron microscope with TIV and could find no morphological difference between the two. SIV has been extensively studied by Day and Mercer (1964) who agree that there is no morphological difference between TIV and SIV; they also find that both viruses have similar sedimentation coefficients, $S_{20} = 2200$.

a. Host Range

The host range seems similar to that of TIV in that it is capable of infecting species of Coleoptera, Diptera, and Lepidoptera. By adding SIV to the medium in which second-instar larvae of the mosquito *Aëdes aegypti* (L.) were developing, Day and Mercer (1964) succeeded in infecting two larvae. A similar experiment carried out previously with TIV in England on the larvae of *Culex* sp. was negative.

An iridescent virus has recently been recorded from the larvae of the mosquito *Aëdes taeniorhynchus* (Wiedemann). From the electron micrographs shown the virus appears morphologically similar to the *Tipula* iridescent virus (TIV) (Clark *et al.*, 1965).

b. Symptomatology and Pathology

There seems to be little difference in the disease response of larvae infected with the two viruses, though there may be some variation in the infection of specific tissues.

c. Isolation of the Virus

In the work carried out at Cambridge the larvae of *Pieris brassicae* L., the large white butterfly, were used in the mass production of large quantities of TIV. Day and Mercer (1964) used larvae of the greater wax moth, *Galleria mellonella* L., for SIV and the following purification technique is quoted from their paper. Infected *Galleria* pupae were macerated in a Teflon grinder in a standard buffer consisting of 0.01 M borate at pH 7.5 containing 0.1 M NaCl. For the preparation of large quantities of virus 0.08% sodium azide was added to the solution to inhibit bacterial growth. These suspensions were clarified at 1000 rpm for 5 minutes in a Servall SS33 rotor. The supernatant was strained through muslin and then centifuged at 15,000 rpm for 20 minutes. A layer of cell debris generally covered the virus pellet, but could readily be removed by gently washing with buffer. The virus pellet was allowed to resuspend in buffer overnight at 4°C, was clarified by low-speed centrifugation, and sedimented again. After resuspending in buffer, each milliliter of virus was added to 22 ml of a sucrose density gradient (40–5% in 0.01 M borate buffer) and centrifuged for 20 minutes at 15,000 rpm in a Spinco SW25 rotor. The visible virus zones were bulked and diluted in buffer, and the virus was sedimented at 15,000 rpm for 20 minutes in a Spinco No. 30 rotor. Pellets were re-

suspended in standard buffer and the virus solution dialyzed against
buffer overnight at 4°C.

d. MORPHOLOGY AND ULTRASTRUCTURE OF THE SIV

At the time of writing there is no information on the ultrastructure of
the SIV particle which seems morphologically identical to the TIV
particle.

e. CHEMICAL AND PHYSICAL PROPERTIES

SIV is thermolabile; infectivity of a dilute SIV solution is destroyed
completely at 60°C for 5 minutes and by 30 minutes at 50°C, but not
by 10 minutes at 50°C. The addition of $MgCl_2$ to a concentration of
1 M in the solution increased the thermal instability of the virus.

SIV is resistant to ether and chloroform and infectivity is retained
after exposure to pH 3.0 and pH 10.5 for 3 hours.

Particle counts of SIV in the electron microscope have been related
to infectivity, to total protein, and to absorbancy at 260 mμ. DNA
content is put at 17.6% which is higher than the figure given for TIV
(12.4%). One absorbancy unit represents 0.044 mg. SIV protein per ml.,
and this is equivalent to 1.8×10^9 infectious units. The total particle/
infective particle ratio is approximately 1 : 1. A single infected *Galleria*
pupa may contain roughly 2 mg. virus protein.

SIV developed in *Galleria* larvae kept at 22°C but not at 28°C
(Day and Mercer, 1964).

The same workers give the amino acid analysis in Table III, assuming
a molecular weight of 28,000.

The writer is indebted to Dr. Alan Bellett for permission to reproduce
some unpublished work on the DNA of the iridescent viruses (Bellett
and Inman, 1966). The preparations have an average molecular weight
of 130 million by sedimentation velocity and electron microscopy, and
110 million by bandwidth in cesium chloride. The largest molecules
are about 73 mμ long, so it seems likely that each virus particle contains
a single molecule of DNA. The DNA has a melting temperature of
80°C, under standard conditions, and a buoyant density of 1.692 ± 0.002
with reference to *E. coli* DNA (1.710). The DNA also shows sharp
"melting" at pH 11.55, and reacts with formaldehyde in 3 minutes at
98°C, but not in 16 hours at 37°C. These properties are consistent
with double-stranded DNA containing about 32% Guanine + Cytosine

TABLE III
AMINO ACID ANALYSIS OF SIV

Amino acid	Number of residues	Amino acid	Number of residues
Lysine	11.7	Alanine	21.9
Histidine	4.0	Half cystine	1.9+
Arginine	18.1	Valine	19.3
Aspartic acid	29.2	Methionine	5.0
Threonine	18.5	Isoleucine	15.5
Serine	26.4	Leucine	19.7
Glutamic acid	16.2	Tyrosine	10.7
Proline	23.9	Phenylalanine	11.0
Glycine	23.7		

and no unusual bases, in good agreement with Thomas's (1961) analysis of TIV. So far no significant differences have been found between DNA preparations from SIV and those from TIV.

There is no doubt that differences exist between TIV and SIV. What is wanted is an equally thorough biochemical investigation of TIV; this might indicate that the similarities between the two viruses are greater than the differences. It would be interesting to propagate SIV for some time in the larvae of *Tipula paludosa* Meig., the original host of TIV, and then compare the properties of the two viruses.

C. Acute Bee Paralysis Virus (ABPV)

All the insect viruses with which we have dealt so far have affected the larval stages with only incidental infection of the adult. We come now to the consideration of two viruses which infect exclusively the adult insect, the honeybee *Apis mellifera* L.

It has long been known that adult bees suffered from a disease called "bee paralysis," with which was associated a rather vague complex of symptoms of which the most reliable was inability to fly and trembling of the wings and legs. Among other symptoms popularly associated with the disease are hairlessness, a dark, greasy appearance, an unpleasant odor, and dysentery. However, Bailey (1965) considers these are not reliable for diagnostic purposes but may depend upon variable factors secondary to infection.

Burnside (1945) was the first to show that a virus was concerned by reproducing the disease in caged healthy bees. He sprayed them with a bacteria-free extract of diseased bees which was successful in nine serial transmissions; he also showed that bees became paralyzed when fed or injected with extracts of diseased bees.

The subject of bee paralysis has been recently investigated by Bailey *et al.* (1963) who discovered that there were two viruses connected with the disease; one, the more virulent of the two, was named "Acute Bee Paralysis Virus" (ABPV) and the other, less virulent, "Chronic Bee Paralysis Virus" (CBPV). Although the natures of the infections by CBPV and ABPV are very similar the viruses themselves are clearly different.

a. Host Range

The following bumblebees have been found susceptible, *Bombus agrorum* (Fabr.), *B. hortorum* (Linn.), *B. lucorum* (Linn.), *B. ruderaris* (Muller), and *B. terrestris* (Linn.). Normal bumblebees, like normal honeybees, may contain small quantities of ABPV (Bailey and Gibbs, 1964).

b. Symptomatology and Pathology

Although there are other causes for bee paralysis besides virus infection, poisoning from the use of insecticides being one, the general picture of the virus disease is fairly typical. Affected bees are unable to fly, and they walk around with abnormal trembling of the legs and wings. Symptoms caused by ABPV are very severe and the bees die quickly. In order to demonstrate unequivocally that ABPV is present in a colony of bees, it is necessary to inject extracts of some bees from the colony into other bees from the same colony using as concentrated an extract as possible. In the histopathology of acute bee paralysis the only abnormalities observed by means of the optical microscope are a strong basophilic reaction in the corpora pedunculata ("mushroom bodies") of the brain, and dense amorphous basophilic deposits in the lumen of the midguts (Bailey and Gibbs, 1964). In acutely paralyzed bees these deposits seem to stream from the midgut epithelium (Bailey, 1965) (Fig. 31).

c. Isolation of the Virus

The following account of the purification of ABPV is taken from the work of Bailey *et al.* (1963). Virus is extracted from bees by grinding

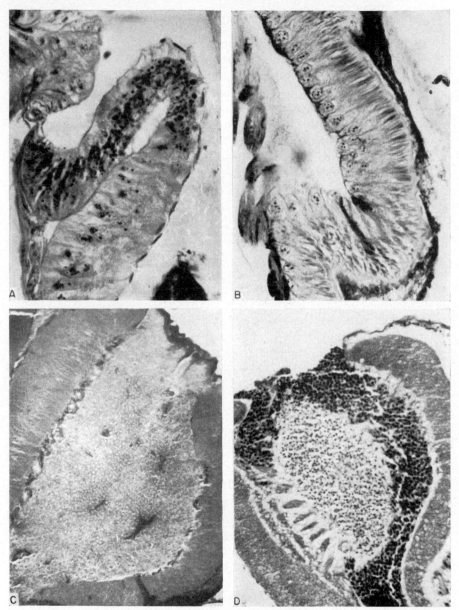

FIG. 31. A and B. Longitudinal sections of the hindgut of bees affected with A, chronic paralysis; and B, acute paralysis. C. Mushroom body of healthy bee. D. Mushroom body of bee with acute paralysis. The darkly, staining deposits in the lumen of the gut are absent in normal bees. Magnification, × 100. (From Bailey, 1965.)

them in tap water and a quarter volume of carbon tetrachloride. The resulting emulsion is coarsely filtered and centrifuged to remove the carbon tetrachloride and any suspended matter including bacteria.

Bees can be infected in one of three ways: by spraying 2.5 ml of a virus suspension into each cage containing about 25 bees; by feeding equal volumes of syrup and virus suspensions to each cage of 20 bees (after 24 hours plain syrup can be substituted); and by injecting bees, anesthetized with carbon dioxide, with 1 ml of a virus suspension with a sterile microsyringe fitted with a 0.3-mm diameter needle. The suspension is injected into the hemocoel through a dorsal intersegmental membrane; only the point of the needle must be inserted to avoid damaging the gut, and it must be kept parallel with and immediately under the cuticle.

To produce enough virus for purification several hundred bees must be infected by injection. Bees injected with ABPV are kept at 30°C for 2 or 3 days; they can then be used immediately or can be frozen and stored at $-20°C$ for use later.

After the virus has been extracted as previously described, it must be purified by one or more cycles of sedimentation and clarification in a high-speed centrifuge. The virus can be sedimented by centrifuging at 75,000 g for 2 hours; it is then resuspended in 0.02 M ammonium acetate (pH 7.0) and clarified by centrifuging at 8,000 g for 10 minutes.

d. MORPHOLOGY OF THE VIRUS

Purified preparations made from acutely paralyzed bees always contain many isometric particles, which in negatively stained preparations measure about 28 mμ in diameter (Fig. 32); about 10^{11} to 10^{12} such preparations can be extracted from each bee. Very few particles, and sometimes none at all, are visible in similar preparations made from naturally paralyzed or healthy bees. Some of the particles seen in negatively stained preparations appear to be empty, as do so many other viruses similarly stained.

The ABPV particles are somewhat similar to another noninclusion virus affecting the cosmopolitan armyworm *Cirphis unipunctata* (Haworth) (see page 102), and also to animal viruses of the picornavirus group (enteroviruses), and many plant viruses.

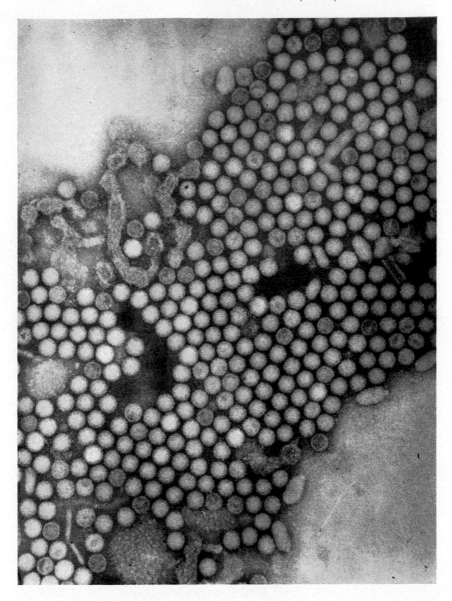

Fig. 32. Electron micrograph of acute bee paralysis virus particles. Magnification, × 240,000. (From Bailey, Gibbs, and Woods, 1964.)

D. Chronic Bee Paralysis Virus (CBPV)

a. HOST RANGE

Similar to that of acute bee paralysis virus (ABPV).

b. SYMPTOMATOLOGY AND PATHOLOGY

Both viruses occur in apparently healthy bees, but only CBPV particles are numerous in diseased bees from colonies naturally infected with the disease known as "bee paralysis." This is evidently the virus which causes the natural disease, since inoculations with it give rise to symptoms very similar to those of natural infections.

In the histopathology of the disease caused by CBPV the cell inclusions described by Morison (1936) are strikingly evident in the cytoplasm of the cells of the hindgut. They are largest in the cells immediately posterior to the openings of the Malpighian tubules and they are strongly basophilic. These inclusions have not been observed in acutely paralyzed bees or in healthy bees.

Unlike the amorphous basophilic deposits in the guts of acutely paralyzed bees, those in the lumen of the gut of chronically paralyzed bees seem cellular and nucleated. They seem to have no obvious connection with the midgut epithelium as appears to be the case with acutely paralyzed bees (Bailey, 1965).

Virus particles, round to oval in shape, have been observed in the thoracic and abdominal ganglia of bees infected with CBPV. The particles appeared either closely packed in inclusions, or free in the cytoplasm of the nerve cells (Lee and Furgala, 1965a).

c. ISOLATION OF THE VIRUS

The purification procedure for CBPV is similar to that for ABPV; bees injected with CBPV are kept for 6 days at 30°C; they can then be used immediately or stored at −20°C. The centrifugation procedure is similar except that the extract must be centrifuged for 2.5 hours at 75,000 g instead of 2 hours as for ABPV.

CBPV can be separated from ABPV by using extracts for each transfer diluted so that each dose contains the equivalent of 10^{-6} to 10^{-8} of a bee. In this way CBPV is transmitted each time while ABPV naturally present in the extracts of the bees is kept too dilute to infect. Since CBPV can be isolated from normal bees, it seems that infection by CBPV is more common than chronic paralysis. Although the virus may be widespread among honey bees, the disease of chronic paralysis seems very localized within countries (Bailey, 1965).

d. MORPHOLOGY OF THE VIRUS

Purified preparations of virus from chronically paralyzed bees contain many irregularly shaped particles, measuring on an average 27×45 mμ, though some may be only 20 mμ wide and others as long as 70 mμ. While ABPV particles resemble some other viruses, the particles of CBPV seem unlike any other virus previously described (Fig. 33) (Bailey et al., 1963), unless, as suggested by Lee and Furgala (1965a), they bear some resemblance to the new type of insect virus described by Vago (1963) (See page 105).

E. Sacbrood Virus (SBV)

This virus attacks the larva and not the adult stage of the honeybee *Apis mellifera* Linn. The name is apparently derived from the saclike appearance of the affected larvae. Spherical particles associated with the disease have been observed in the electron microscope by Steinhaus (1949), by Brčák et al. (1963), and in sections of diseased tissue by Lee and Furgala (1965b). That these particles were the actual virus was first demonstrated by Bailey et al. (1964), by means of infection tests.

a. HOST RANGE

Not much is known of the biology of sacbrood but at the moment the virus has only been recorded from the larvae of the honeybee.

b. SYMPTOMATOLOGY AND PATHOLOGY

Externally affected larvae are light brown in color with a toughened cuticle. In late stages of the disease the body contents become liquefied in a manner similar to the contents of larvae affected with the nuclear polyhedroses.

There is very little information on the pathology of the disease. Lee and Furgala (1965b) demonstrated the presence of isodiametric particles 28 mμ in diameter in infected tissue of diseased honeybee larvae. The exact situation was not determined though it was thought to be the cytoplasm of fat-body cells. The particles were either randomly distributed or arranged in crystalline array (Fig. 34), somewhat after the same manner as TIV (Fig. 24).

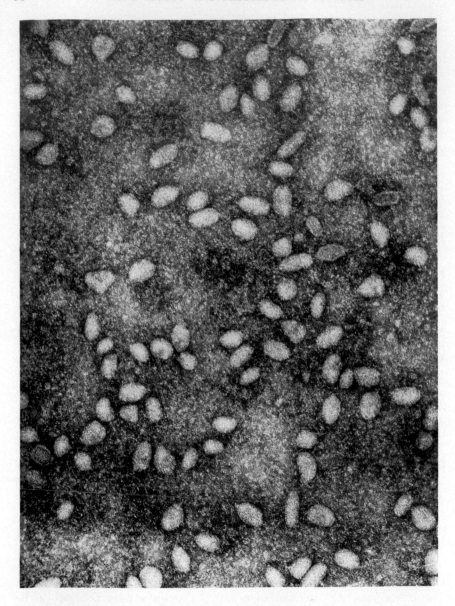

FIG. 33. Electron micrograph of chronic bee paralysis virus particles. Magnification, × 240,000. (From Bailey, Gibbs, and Woods, 1963.)

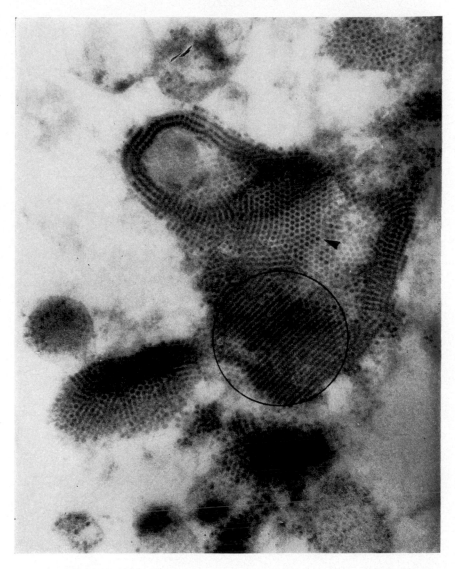

Fig. 34. Sacbrood virus with large inclusion body lying free in the cytoplasmic matrix. Arrow points to particles forming isodiametric shape. Particles in highly compacted arrays form a crystal lattice with linear pattern and periodic spacing of 30 mμ (circle). Magnification, × 60,000. (From Lee and Furgala, 1965b.)

Young larvae are more susceptible to infection than older larvae; in experiments carried out by Bailey *et al.* (1964) 80% of larvae 2–3 days old developed sacbrood whereas only 40% of larvae 4–5 days old did so. The LD_{50} of SBV for 2-day-old larvae was 10^5 to 10^6 particles.

c. Isolation of the Virus

The preliminary steps in purification are similar to those used in the purification of ABPV. After this partial purification and concentration by centrifugation, SBV can be further purified by centrifugation in a sucrose gradient.

d. Morphology and Ultrastructure of SBV

In shadowed preparations Brcak *et al.* (1963) showed the virus particles to be apparently spherical and to measure about 30 ± 2 mμ.

Particles of SBV stained with phosphotungstic acid measure 28 mμ in diameter (Fig. 35). The particles closely resemble those of ABPV both in size and appearance and both have a sedimentation constant (S_{20}) of about 160. Though the particles of ABPV and SBV share some physical properties, no relationship between them was suggested by infectivity or serological tests. Attempts to infect larvae with ABPV by feeding them high-titer extracts, or to infect adult bees by injecting them with SBV have all proved negative. It is interesting that the two viruses ABPV and SBV should apparently be specific for different stages of the same insect (Bailey *et al.*, 1964). Some information on the ultrastructure of the SBV particle has been obtained by Brcak and Králík (1965). Using negative staining and the photographic rotation technique of Markham *et al.* (1963), they established the axis symmetry 5 : 3 : 2, and distinguished twelve subunits on the periphery of the particle. The original (nonrotated) picture suggested that one edge is formed from three spherical subunits. Brčák and Králík conclude that the surface of the particle is an icosahedron formed from 42 subunits.

F. Wassersucht Virus of Coleopterous Insects

There are not many instances of virus diseases of Coleoptera recorded so far. Smith *et al.* (1961) experimentally infected the larvae of *Tenebrio molitor* Linn., *Agriotes obscurus* Linn., and *Melolontha* sp. with the *Tipula* iridescent virus (TIV). Steinhaus and Leutenegger (1963) recorded the natural infection of a scarab larva *Sericesthis pruinosa*

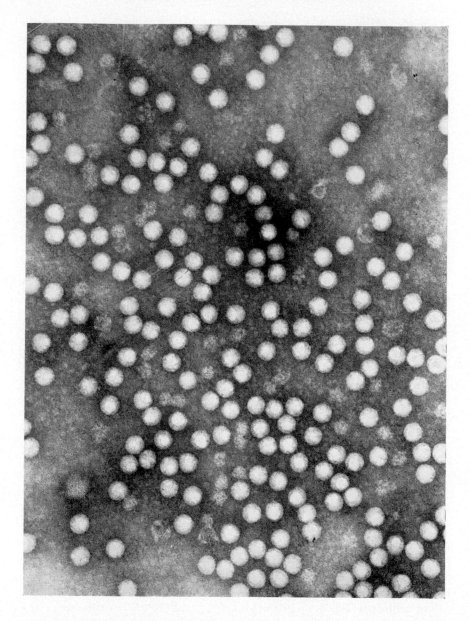

FIG. 35. Sacbrood virus, negatively stained with phosphotungstic acid. Magnification, × 240,000. (From Bailey, Gibbs, and Woods, 1964.)

(Dalman) with a similar iridescent virus (SIV). A new virus from the larva of *Melolontha melolontha* has recently been recorded (Hurpin and Vago, 1963).

Heidenreich (1939) described a disease which he considered to be due to a virus in the larvae of *Melolontha melolontha* Linn. and *M. hippocastani* Fabr. He called it "Wassersucht," or watery degeneration, because of the appearance of the diseased larvae. Surany (1960) described apparently similar diseases in larvae of tropical species of the same order. He called them the histolytic disease and Heidenreich's disease. What is possibly the same disease was described as "maladie transparente" of *Melolontha* sp. (Hurpin and Vago, 1958).

A more intensive study of the Wassersucht disease has been made by Krieg and Huger (1960a,b) and the following account is derived from their work.

a. Host Range

At the time of writing the virus has been recorded only from various species of *Melolontha, M. melolontha* Linn. and *M. hippocastani* Fabr. What may be the same disease occurs in the larvae of the tropical species known as rhinoceros beetles.

b. Symptomatology and Pathology

The outstanding symptom seems to be the transparency of the larva, especially in the abdomen; this is due to the watery disintegration of the fat body. In the late stages of the disease the larvae show irregular locomotion probably as a result of a loss of coordination. As the disease progresses dehydration sets in combined with a striking shrinkage; after death the larvae become mummified without visible putrefaction.

The first internal symptom develops in the fat body, in which the albuminoid spheres, very plentiful in normal insects, show a marked degradation. The progressive stages of this degradation can be followed via the light microscope and are of diagnostic importance. According to Krieg and Huger these albuminoid spheres are converted into virogenic stroma consisting of granular elements corresponding to incomplete virus particles. This matter is dealt with further in Chapter VI dealing with the multiplication of viruses.

The different types of hemocytes react differently to infection. The plasmatocytes partly undergo vacuolate degeneration, whereas the spherule cells are not affected. At first, the epidermis, the tracheal system, the muscles, the nervous system, and the intestinal wall remain intact. As the disease progresses these unaffected tissues become atrophied.

The virus appears to be confined to the cytoplasm and in this respect resembles TIV, but differs from it in containing RNA. Attempts to infect larvae by feeding have been unsuccessful, but injection of the virus was quite effective. On this account Krieg and Huger suggest the virus may be transmitted through the egg.

In their sections of tissue examined in the electron microscope some free virus particles were observed and also a "network" containing virus particles which could be analagous to the virogenic stroma of other insect virus diseases.

c. Isolation of the Virus

The virus particles were isolated by gradient centrifugation of the hemolymph.

d. Morphology of the Virus Particles

The particle appears to be near-spherical and measures between 60 and 75 mμ in diameter (Krieg and Huger, 1960b).

An interesting virus of a new type which affects the larvae of the Indian rhinoceros beetle, *Oryctes rhinoceros* Linn., has been described by Huger (1966).

Larvae affected by this "Malaya disease" exhibit characteristic symptoms; the abdomen shows a turbid, glassy, almost pearly appearance and diarrhea is common. Towards the end of the disease, the larvae appear shiny, beige-colored, and waxy, especially in the abdominal region which is often broader than normal. At this juncture there is frequently a development of chalky, white, more or less concrete, bodies under the larval integument, especially in the abdominal region.

The virus is easily transmissible to third-instar larvae both by contaminated food and intrahemocoelic injection.

The virus seems to multiply chiefly in the nuclei of fat-body cells, but virogenic stromata also occur in the cytoplasm. There appear to be two forms of the virus, "spherical" particles with polygonal contours of about 164 mμ diameter, and rod-shaped particles measuring about 200×70 mμ. In an advanced stage of the disease the hypertrophied nuclei contain large numbers of single- or double-membraned "spherical" particles, and, closely associated with them, masses of the virus rods, which may form pseudocrystalline patterns. The virus rods have a highly differentiated structure and are considered to be the mature form of the virus. They are said to arise (1) from dense chromatin-like virogenic stroma; (2) from loose, fibrillar, and granular stromal material; (3) from filaments; and (4) from inside the single- or double-membraned spherical particles (shells). These last are thought to represent viral developmental stages.

There are some points of resemblance in the symptoms of this disease to those of the Wassersucht disease of *Melolontha,* just described, of Heidenreich's disease of *Oryctes* (Surany, 1960), and of the "maladie transparente" of *Melolontha* sp. (Hurpin and Vago, 1958).

G. A Virus from the Armyworm *Cirphis unipuncta* (Haw.)

A noninclusion virus has been isolated from the armyworm *Cirphis unipuncta* (Haw.) (Steinhaus, 1951; Wasser, 1952). Not very much is known about this virus, and it is important chiefly as being the first virus of this type to be isolated from an insect and to be characterized in the electron microscope.

Larvae of *C. unipuncta* infected in late third instar soon appear swollen and somewhat darker than normal insects. The cuticula of the diseased larvae have a waxy appearance, and in some cases the midportion is slightly enlarged. The liquefaction and disintegration of tissues characteristic of nuclear polyhedroses are absent. Deaths from the disease may occur in the larval or pupal stage; the greatest percentage of mortality occurs in the pupal stage when mortalities of 90 to 100% are common. As the disease progresses, the larvae become sluggish and cease to feed; death ensues 6–14 days after infection.

Larvae are readily infected *per os* and the virus may be highly virulent.

The virus particles are small, measuring about 25 mμ in diameter; they appear to be approximately spherical.

H. A Virus from *Antheraea eucalypti* Scott

a. Host Range

This is a newly recorded virus from the saturniid *Antheraea eucalypti* Scott, and so far it has only been found in this species. It has been described by Grace and Mercer (1965) and the following information is derived from their work. The suggestion is made that this virus should be called a "nonoccluded" virus since the occurrence of particles in dense masses gives rise to "inclusions" visible in the optical microscope. However, the same can be said of other viruses of this group, notably the *Tipula* iridescent virus. In the context of this book, therefore, "noninclusion" virus means that the virus particles are not occluded in a protein crystal as in the polyhedroses and granuloses.

b. Symptomatology and Pathology

Diseased larvae are discolored and flaccid, usually with a constriction of the body towards the hind end; they sometimes hang by the last pair of prolegs. The onset of the disease is quite rapid and invariably fatal; larvae which were feeding and appeared normal would be dead within 12 hours. When moribund larvae are dissected, the midgut appears to be the only organ obviously affected; the cells are completely lysed and little remains except the peritrophic membrane, packed with food. The hindgut usually contains decomposing fecal pellets which apparently cannot be voided.

Sections of diseased cells of the midgut stained with toluidine blue or hematoxylin and eosin show large intranuclear, inclusion bodies. Electron micrographs of sections of the diseased midgut revealed the presence of the virus in enormous numbers (see TIV page 77). They develop in the nucleus where extensive close-packed arrays of particles form the inclusion bodies visible in the optical microscope. Eventually the nuclear membrane disappears, and particles are then found in the cytoplasm. Finally the cell membrane of the gut breaks down and the particles are released.

c. Isolation of the Virus

The following purification method was used by Grace and Mercer (1965).

Whole dead larvae or their extracted guts were macerated in water and the coarser fragments sedimented. The supernatant was further purified by several centrifugations at increasingly high speeds (from 2000 to 39,000 rpm). The pellet obtained after centrifugation at 39,000 rpm was resuspended in 1–2 ml of water and placed on a sucrose density gradient (5–40% sucrose). In this gradient a discrete band was obtained after 60 minutes centrifugation in a swing-out (Spinco SW25) head at 24,000 rpm. The band was removed and dialyzed for 24 hours against water.

d. MORPHOLOGY AND ULTRASTRUCTURE OF THE VIRUS

The virus particle appears spherical and measures 50 mμ in diameter; it contains a core 30 mμ in diameter which appears to stain more deeply with PTA. It is interesting that the stain is apparently able to enter the intact particle and delineate the core. These cores seem to be liberated on occasion from the particle, possibly by the action of the PTA. The surface coat is composed of a number of capsomeres which measure about 80–100 Å, but the exact number is not yet known. The virus particle somewhat resembles the polio virus but is rather larger; it is also similar in appearance to the acute bee paralysis virus (ABPV).

I. *Drosophila* σ Virus

a. HOST RANGE

The virus has been transferred by artificial inoculation to *Drosophila simulans; D. wellistoni, D. prosaltans, D. gibberosa,* and *D. funebris.*

b. SYMPTOMATOLOGY

Among natural populations of *Drosophila melanogaster,* some individuals are readily found which show a highly aberrant behavior when brought into contact with carbon dioxide. They are called CO_2-sensitive flies; resistant flies can be kept for hours in pure CO_2 without any permanent injury. The behavior of sensitives is strikingly different; within a broad range of temperature and CO_2 pressure, they are unable to recover from even a 30-second contact with the gas. Carbon dioxide sensitivity was first discovered in a laboratory strain of *Drosophila* by L'Heritier and Teissier in 1937, and a comprehensive review of the subject was written by L'Heritier in 1958, from which most of this information is derived.

The virus, now known as *virus* σ, can be introduced into its host either through artificial inoculation or through inheritance by any one of the two gametes. In every case, its presence in the fly is not harmful and brings about only one detectable symptom, CO_2 sensitivity. Sensitive flies seem to live as long and to lay as many eggs as do resistant ones. Nothing is known of the intimate physiological mechanism underlying sensitivity, but L'Heritier (1948) discusses a specific action of CO_2 on the motor nerve cells.

Two very different kinds of sensitives have been described; these are based on the rate of multiplication of the virus and its ability to be inherited.

They are known respectively as *stabilized* and *nonstabilized* types. The stabilized females consistently transmit the virus to all their progeny. The nonstabilized females are in the majority and transmit the virus to about 22% of their progeny and that only for a limited period.

c. The Virus

A fairly pure infectious supernatant can be prepared in the following way. Flies are crushed in a 30% sucrose solution buffered with phosphate. Coarse debris is eliminated by a first centrifugation at 3000 rpm, followed by two successive centrifugations of the supernatant at 18,000 rpm. Without the sucrose the virus would be inactivated by this procedure, but the density increase prevents any significant fall in titer. The sucrose is then removed by dialysis in a cellophane bag immersed in cold Ringer's solution. This procedure gives a clear, colorless preparation equal in titer to the first crude extract.

It is probable that the σ virus could now be isolated by density gradient centrifugation using either sucrose or cesium chloride (Brakke, 1951).

Since it has not been possible, so far, to isolate the virus, there is no information about it from the point of view of its electron microscopy. Some indirect information on its size has been obtained by means of X-ray irradiation and ultrafiltration through gradocol membranes. From these techniques the results suggest a size between 30 and 40 mμ.

PART II. MISCELLANEOUS VIRUS DISEASES

A. A Virus from *Melolontha melolontha* (Linn.)

A recently discovered virus which affects the larvae of the beetle *Melolontha melolontha* Linn. does not fit into any of the categories of insect viruses so far described, although it more nearly approaches the cytoplasmic polyhedroses. For the time being, it is dealt with separately in a category of its own. The disease was first observed by Hurpin and Vago (1963) who gave a preliminary description of the affected larva and the intracellular inclusions.

a. Host Range

The only full description available at present is of the disease as it attacks *Melolontha melolontha* Linn., but Weiser (1965) mentions a similar infection in the larva of *Acrobasis* sp. (Lepidoptera), and another in the larvae of some Diptera.

b. Symptomatology and Pathology

Diseased larvae are chalky white and the fat body is slightly distended. On dissection, the fat body has a somewhat foamy appearance. Only the fat body appears to be infected, and infection is limited largely to the central part and to isolated fat cells lining the hypodermal layer. Crystalline intracellular inclusions are formed and occur only in these cells; the normal fat droplets are reduced in number and in heavily infected cells are replaced by the crystals. It is not correct to refer to these as polyhedra since they are actually spindle-shaped, but they contain the virus particles in a similar manner to polyhedra.

In the microscope, the first sign of the disease is the development of small rhomboidal crystals, 0.7–1 μ long, located in the nodi between fat droplets. At this stage they stain with aniline dyes; later they may attain a size of 13 μ long by 6 μ broad, and then become resistant to staining. Usually one crystal per cell broadens to an oval body, 12 μ broad and 19–20 μ long. In it a cluster of needlelike refringent bodies can be seen, and at this stage virus particles are visible inside the crystal (Weiser, 1965).

c. Morphology and Ultrastructure of the Virus

Sections of the large type of inclusion body revealed the presence of round to oval, laterally concave bodies with a diameter ranging from 250 to 40 mμ, often in regular lines (Vago, 1963). This regularity of arrangement of the virus particles inside the crystal is similar to that found in the nuclear polyhedra of *Tipula paludosa* Meig. (see page 31).

Vago and Croissant (1964) used the following method to obtain electron micrographs of the virus. A minute fragment of infected fat body is shaken up in 0.25 ml of a 2% solution of phosphotungstic acid at pH 7.4; a microdrop of the solution is then placed on the microscope grid on which is a carbon film.

The virus particle appears to be very large, oval to round in shape, and measuring 370 × 250 mμ; the surface is composed of a number of regularly placed subunits.

This virus seems to be unlike any of the other insect viruses so far described and resembles more clearly some of the larger viruses affecting vertebrate animals.

B. A Suspected Noninclusion Virus in the European Corn Borer *Ostrinia nubilalis* (Hübner)

A condition resembling a virus disease in the European corn borer *Ostrinia nubilalis* (Hübner) was reported by Raun (1963). In affected larvae the integument became opaque, and the fat-body cells showed pyknotic nuclei and cell-wall breakdown. Dying larvae became dehydrated and finally mummified.

Adams and Wilcox (1965) have made a preliminary electron microscope study of sections of diseased fat body in the same insect. Hexagonal particles, possibly icosahedra, about 60 mμ in diameter were observed in the cytoplasm of the cells. Also present in the fat-body cells were possible virogenic stromata; the albuminoid spheres showed a breakdown somewhat similar to that described by Krieg and Huger (1960b) in a noninclusion virus disease in the larvae of *Melolontha melolontha*.

REFERENCES

Adams, J. R., and Wilcox, T. A. (1965). A viruslike condition in the European corn borer, *Ostrinia nubilalis* (Hübner). *J. Invert. Pathol.* **7**, 265–266.

Anderson, E. S., Armstrong, J. A., and Niven, J. S. F. (1959). Fluorescence microscopy: Observations of viruses growing with amino-acridines. *Symp. Soc. Gen. Microbiol.* **9**, 224–255.

Anderson, T. F. (1952). Stereoscopic studies of cells and viruses in the electron microscope. *Am. Naturalist* **86**, 91.

Bailey, L. (1965). Paralysis of the honey bee, *Apis mellifera* (Linn.). *J. Invert. Pathol.* **7**, 132–140.

Bailey, L., and Gibbs, A. J. (1964). Infection of bees with acute paralysis virus *J. Insect Pathol.* **6**, 395–407.

Bailey, L., Gibbs, A. J., and Woods, R. D. (1963). Two viruses from adult honey bees (*Apis mellifera* Linn.). *Virology* **21**, 390–395.

Bailey, L., Gibbs, A. J., and Woods, R. D. (1964). Sacbrood virus of the larval honey bee (*Apis mellifera* Linn.). *Virology* **23**, 425–429.

Bellett, A., and Inman, D. (1966). In press.

Brakke, M. K. (1951). Density gradient centrifugation: a new separation technique. *J. Am. Chem. Soc.*, **73**, 1847.

Brakke, M. K. (1960). Density gradient centrifugation and its application to plant viruses. *Advan. Virus Res.* **7**, 193–224.

Brčák, J., and Králík, O. (1965). On the structure of the virus causing sacbrood of the honey bee. *J. Invert. Pathol.* **7**, 110–111.

Brčák, J., Svoboda, J., and Králík, O. (1963). Electron microscopy investigation of sacbrood of the honey bee. *J. Insect Pathol.* **5**, 385–389.

Brenner, S., and Horne, R. W. (1959). A negative staining method for high resolution electron microscopy of viruses. *Biochim. Biophys. Acta* **34**, 104–110.

Burnside, C. E. (1945). The cause of paralysis of bees. *Am. Bee J.* **85**, 354–355.

Clark, T. B., Kellen, W. R., and Lum, P. T. M. (1965). A mosquito iridescent virus (MIV) from *Aëdes taeniorhynchus* (Wiedemann) *J. Invert. Pathol.* **7**, 519–520.

Day, M. F., and Mercer, E. H. (1964). Properties of an iridescent virus from the beetle, *Sericesthis pruinosa*. *Australian J. Biol. Sci.* **17**, 892–902.

Grace, T. D. C., and Mercer, E. H. (1965). A new virus of the Saturniid *Antheraea eucalypti* Scott. *J. Invert. Pathol.* **7**, 241–244.

Heidenreich, E. (1939). *Proc. 7th Intern. Congr. Entomol., Berlin, 1938* S, 1963.

Hills, G. J., and Smith, K. M. (1959). Further studies on the isolation and crystallization of insect cytoplasmic viruses. *J. Insect Pathol.* **1**, 121–128.

Huger, A. (1966). A virus disease of the Indian rhinoceros bettle *Oryctes rhinoceros* Linn., caused by a new type of insect virus, *Rhabdionvirus oryctes*. gen. n. sp. n. *J. Invert. Pathol.* **8**, 38–51.

Hurpin, B., and Vago, C. (1958). Les maladies du Hanneton commun (*Melolontha melolontha*) *Entomophaga*, **3**, 285–330.

Hurpin, B., and Vago, C. (1963). Une maladie a inclusions cytoplasmiques fusiformes chez le coleoptère *Melolontha melolontha* L. *Rev. Pathol. Vegetale Entomol. Agr. France* **42**, 115–117.

Kaesberg, P. (1956). Structure of small 'spherical' viruses. *Science* **124**, 626–628.

Krieg, A., and Huger, A. (1960a). A virus disease of coleopterous insects. *J. Insect Pathol.* **2**, 274–288.

Krieg, A., and Huger, A. (1960b). Uber eine Viruskrankheit bei Coleopteren. *Naturwisserschaften* **17**, 403–404.

Lee, P. E., and Furgala, B. (1965a). Chronic bee paralysis virus in the nerve ganglia of the adult honey bee. *J. Invert. Pathol.* **7**, 170–174.

Lee, P. E., and Furgala, B. (1965b). Electron microscopy of Sacbrood virus *in situ*. *Virology* **25**, 387–392.

L'Heritier, P. (1948). Sensitivity to CO_2 in *Drosophila*: A review. *Heredity* **2**, 325–348.

L'Heritier, P. (1958). The hereditary virus of *Drosophila*. *Advan. Virus Res.* **5**, 195–244.

L'Heritier, P., and Teissier, G. (1937). *Compt. rend.* **205**, 1099.

Markham, R., Frey, S., and Hills, G. J. (1963). Methods for the enhancement of image detail and accentuation of structure in electron microscopy. *Virology* **20**, 88–102.

Morison, G. D. (1936). Bee paralysis. *Rothamsted Conf.* **22**, 17–21.

Raun, E. S. (1963). A virus-like disease of the European corn borer. *Proc. N. Central Branch Entomol. Soc. Am.* **18**, 21.

Smith, K. M. (1955). What is a virus? *Nature* **175**, 12–14.

Smith, K. M., and Hills, G. J. (1959). Further studies on the electron microscopy of the *Tipula* iridescent virus. *J. Mol. Biol.* **1**, 277–280.

Smith, K. M., and Hills, G. J. (1962). Ultrastructure and replication of insect viruses. *Proc. 5th Intern. Conf. Electron Microscopy, Philadelphia, 1962* Vol. II, Art. V-1. Academic Press, New York.

Smith, K. M., Hills, G. J., and Rivers, C. F. (1961). Studies on the cross-inoculation of the *Tipula* iridescent virus. *Virology* **13**, 233–241.

Steinhaus, E. A. (1949). Nomenclature and classification of insect viruses. *Bacteriol. Rev.* **13**, 203–223.

Steinhaus, E. A. (1951). Report on diagnoses of diseased insects 1944–1950. *Hilgardia* **20**, 629–678.

Steinhaus, E. A., and Leutenegger, R. (1963). Icosahedral virus from a scarab (*Sericesthis*). *J. Insect Pathol.* **5**, 266–270.

Surany, P. (1960). Diseases and biological control in rhinoceros beetles. *South Pacific Comm. Tech. Paper* **128**, 1–62.

Thomas, R. S. (1961). The chemical composition and particle weight of the *Tipula* iridescent virus. *Virology* **14**, 240–252.

Thomas, R. S., and Williams, R. C. (1961). Localization of DNA and protein in *Tipula* iridescent virus (TIV) by enzymatic digestion and electron microscopy. *J. Biophys. Biochem. Cytol.* **11**, 15–29.

Vago, C. (1963). A new type of insect virus. *J. Insect Pathol.* **5**, 275–276.

Vago, C., and Croissant, O. (1964). Etude morphologique du virus de la maladie a fuseaux des coleoptères. *Entomophaga* **9**, 207–210.

Wasser, H. B. (1952). Demonstration of a new insect virus not associated with inclusion bodies. *J. Bacteriol.* **64**, 787–792.

Weiser, J. (1965). *Vagoiavirus* gen. n., a virus causing disease in insects. *J. Invert. Pathol.* **7**, 82–85.

Williams, R. C., and Smith, K. M. (1957). A crystallizable insect virus. *Nature* **179**, 119–120.

Williams, R. C., and Smith, K. M. (1958). The polyhedral form of the *Tipula* iridescent virus. *Biochim. Biophys. Acta* **28**, 464–469.

Xeros, N. (1954). A second virus disease of the leatherjacket, *Tipula paludosa*. *Nature* **174**, 562.

CHAPTER VI

Mode of Replication of Insect Viruses

Introduction

There is much confusion and little exact knowledge at the present time on the replication of the insect viruses. With the development of insect-tissue culture techniques, however, it should soon be possible to remedy this state of affairs.

The idea of a "life-cycle" was first put foward by Bergold (1950) who made an examination of virus suspensions under the electron microscope, and arranged a sequence of the particles found into a developmental and multiplication cycle. He then proposed the following cycle for the rod-shaped polyhedrosis and granulosis viruses. The development of these viruses begins with one or several small spheres (measuring about 20 mμ in diameter) that grow within a "developmental" membrane. These spheres elongate to kidney and then to V-shaped forms, and appear finally as straight rods still within the developmental membrane. At some period during this process each rod becomes surrounded by the intimate membrane. The rods with both membranes constitute the infectious virus that attaches to the host cell and nuclear membrane by a thin protrusion. In an unknown way spherical subunits are released from the intimate membrane; these spheres begin the cycle again (Bergold, 1958).

Bird (1959) gives his interpretation, and suggests the following developmental sequence. The polyhedra are ingested by a larva and dissolve in the gut, liberating virus rods. The rods attach to the midgut cells and release infectious subunits. These pass through the midgut epithelium, which, in Lepidoptera, is not susceptible to infection, and

110

enter the nucleus of a susceptible cell. Multiplication takes place within the nucleus, and spherical bodies are formed; each sphere contains "germs" for a number of virus particles, the mature form of which is rod-shaped. Bundles of rods, still contained by their developmental membranes, are surrounded by protein which crystallizes to form polyhedral inclusion bodies. Rods from some of the bundles escape from their developmental membranes, are never occluded by polyhedral protein, and attach to other materials in the nucleus. They release subunits which penetrate and pass out through the nuclear membrane and infect adjacent cells.

These conceptions of Bergold and Bird are plainly influenced by ideas of the conventional development of microorganisms and seem out of step with the biosynthesis of other viruses. Furthermore, the "subunits" of the rods, and the "germs" have never been seen in the electron microscope.

Figure 5 illustrates the virus from the nuclear polyhedrosis of *Bombyx mori* Linn., the silkworm, after treatment with weak alkali. In the consequent breakdown of the virus rods the first stage in the liberation of the contents of the intimate membrane is the peeling off of the outer membrane. The membrane breaks in the center and folds backwards, thus forming two spheres still joined in the middle. These finally break apart and are thought to be the same as Bergold's small spheres which he said were discharged from the intimate membrane (Bergold, 1958).

It is quite clear as was first stated by Smith and Xeros (1954) that the virus rod with its intimate membrane is formed first and afterwards enveloped by the outer ("developmental") membrane, and not the contrary as stated by Bergold. Indeed in a recent paper, Bird (1964) admits this, and states furthermore that the theory that viral development begins with one or several bodies growing within a membrane is based on artifacts.

A. Nuclear Polyhedrosis Viruses

a. LATENT PERIOD

The latent period, or the delay in the development of symptoms after inoculation, has been studied by a number of workers. It is influenced by temperature and the size of the dose of virus. Silkworms showed no symptoms 15 days after infection at temperatures of 16° and 17°C, whereas symptoms developed after 6 days at 25°C (Paillot, 1930). In

Colias eurytheme Boisduval, the alfalfa caterpillar, the incubation period varies from an average of 7 days in the summer to 3 weeks in the winter season.

Aizawa (1953) examined the effect of temperature and size of inoculum upon the duration of the latent period in silkworm pupae. When the temperature is constant, the lower the virus concentration the longer is the latent period. If the virus concentration is constant, the higher the temperature the shorter is the latent period. The latent period of virus-injected pupae was from 30 to 36 days at 10°C and no symptoms were observed after 36 days at 5°C.

Krieg (1955) also found that the latent period varied according to the size of the dose and the temperature following virus inoculation. In the nuclear polyhedrosis of the European pine sawfly, *Neodiprion sertifer* (Geoffroy), the LT_{50} (median lethal time), after inoculation with 10^6, 10^4, and 10^2 polyhedra per milliliter was 7, 8.5, and 10 days respectively. With the same virus dose but at temperatures of 29.6°C, 20.8°C, and 11.5°C, the LT_{50} was 4.5, 8, and 21 days.

According to Ayuzawa (1961) the latent period may vary according to the strain of silkworm being tested.

b. Virus Multiplication as Measured by Infection Experiments

Aizawa (1959) estimated the amount of virus in the supernatant of silkworm hemolymph after virus injection. He considered that there was a decrease phase, an increase (logarithmic) phase, and a stationary phase in the LD_{50} time curve. When the temperature was lowered after virus injection the gradient of the increase phase became less steep. The decrease phase could be due both to an eclipse phase and to adsorption of the virus onto cells.

An eclipse period has also been postulated by Yamafuji *et al.* (1954) and by Krieg (1958).

c. The Morphology of Virus Replication

The development of the nuclear polyhedrosis virus in the cell nuclei of the silkworm, *B. mori*, at various stages of infection has been studied by Smith and Xeros (1953). The study arose out of the discovery of a peculiar structure called the "nuclear net" (Smith *et al.*, 1953), and the sequence of events seems to be briefly as follows. At first, prior to the formation of the nuclear net, the nucleus becomes greatly enlarged; next, the chromatin bodies begin to clump together to form the nuclear net, but at this stage no virus particles are visible. A little later the virus rods begin to form and can be seen protruding from the central chromatic net and from the peripheral chromatin, there being an apparent metamorphosis of the virus rods from the material of the chromatin. They are then pushed out into the ring zone of the nucleus. Following this, many free virus rods accumulate in the nuclear ring, but there are no outer membranes enveloping the rods at this stage; these are secondarily deposited round them in the ring zone. The earliest stage at which free virus rods have been found is just after the formation of the central chromatic mass. At no earlier stage have rods been found nor have they ever been detected in the cytoplasm. On occasion it seems as if several bundles of rods are protruded from the central chromatic mass and from the peripheral chromatin bodies into the ring zone where the individual rods separate from one another. It appears as if the length of the virus rod may not be constant; a long virus rod, 500 μ in length, has been seen protruding from peripheral chromatin. Similar rods almost twice the normal length are sometimes present, giving the appearance of breaking across to form smaller-sized rods. These rods of abnormal length are somewhat reminiscent of the long rods described in the ensuing section on the granulosis viruses.

The distribution of the virus rods in the infected nuclei show several features which suggest that they form in the central chromatic material. A great proportion of the rods, particularly while they are still few in number, are found partially embedded in the chromatic material, presumably in the process of slipping out of these masses (Smith and Xeros, 1954).

What was presumably the nuclear net has been studied further by Xeros (1955, 1956) under the name of a "virogenic stroma." The proteinaceous virogenic stromata form *de novo* in the nuclear sap of infected cells; they become increasingly proteinaceous and Feulgen-positive as they grow and develop. Morphologically they are networks, and virus rods differentiate within vesicles in their cords. According to Xeros these begin as fine rodlets about 60 Å $\times$ 1200 Å in size and increase *in situ* to their final size of 280 Å $\times$ 2800 Å. They are then set free from their vesicles into the pores of the net by disruption of the surrounding cord material and may ultimately reach the ring zone between the centrally placed virogenic mass and the nuclear membrane. The freed virus rods then become enveloped by independently formed membranes as previously stated.

In an electron microscope study of the virus rods it was shown that at very high magnification the contents of the intimate membrane give the appearance of a widely spaced helix; these contents are discharged from either end of the intimate membrane and appear to uncoil as they flow out (Fig. 5). It is thought that this helix is in part DNA and it differs markedly from Bergold's, "subunits."

Although the actual assembly of the parts joining a new infective rod has not been observed, the presence of a helical structure similar to that observed in the rod of the tobacco mosaic virus strongly suggests a similar method of assembly (Smith and Hills, 1962).

Krieg (1961) has also made a study with the electron microscope of the virus rods from nuclear polyhedroses and considers that they show a helical structure. Based on his former observations of a central hole in the subunits he has proposed a model for rod-shaped viruses of insects similar to that made by Franklin *et al.* (1959) for tobacco mosaic virus.

B. Cytoplasmic Polyhedrosis Viruses

Little investigation has up to the present been made into the replication of this type of virus. Because they are mostly small or very small icosahedra it is a difficult problem to follow the details of replication. Xeros (1956) has studied a natural infection of half-grown larvae of the armyworm *Thaumatopoea pityocampa*. In this disease he found virogenic stromata analogous to those formed in the nuclear polyhedroses. They arise subapically in the infected cells. Each virogenic stroma is a definite structural entity, and in a damaged cell survives as a discrete body when the cytoplasm has been destroyed. The stromata have a micronet structure in the cords of which the virus particles arise, at first particularly in the more superficial parts of the masses. When the virus particles have been formed the cord material around them disrupts and liberates them into the larger pores formed as a result of the dissolution of the cords.

The virogenic stromata grow considerably (up to 10 $\mu \times 2$ μ) before small polyhedra, less than $\frac{1}{2}$ μ in diameter, form, distributed irregularly over their surfaces. The freed virus particles become occluded in the polyhedra.

The cytoplasmic virogenic masses, unlike those of the nuclear polyhedroses, are composed of protein and ribonucleic acid, and no deoxyribonucleic acid has ever been detected in them by Feulgen staining.

C. Granulosis Viruses

The replication of the rod-shaped particles of the granulosis viruses seems to be essentially the same as that of the virus particles in the nuclear polyhedroses and a similar helical structure has been observed (Smith and Hills, 1962) (Fig. 21).

An interesting and unusual phenomenon is associated with the granuloses, the significance of which is not fully understood.

There are connected with the diseases large numbers of very long virus rods or threads which are always present. Their presence is easily demonstrated in the following manner. When a lepidopterous larva such as a cutworm in an advanced stage of granulosis is macerated in distilled water as much as 10 ml of a fluid of the consistency of thin cream is obtained. This consistency is due to the suspension of vast numbers of capsules, each of which contains, as we have already seen, a single short virus rod. Centrifugation of this fluid for 30 minutes at 5000 rpm sediments the capsules, leaving a greenish opalescent supernatant. Examination of this supernatant in the electron microscope shows large numbers of the long virus threads, many of them branched in an intricate and bizarre fashion. Also occurring in the supernatant are many empty capsules, truncated to give a characteristic U-shape. An electron microscope study of these long threads and their concomitant capsules suggested the following sequence of events. Negative staining with PTA showed the occluded rod apparently emerging from the capsule; in doing so it pushes off a detachable lid or cap. This gives the empty capsule the characteristic truncated shape of a capital "U" already mentioned (Fig. 36). The socket which contained the short virus rod can be plainly seen; the outer membrane which enveloped the rod is sometimes left behind and sometimes pushed out by the emerging rod. Once it is free in the cell cytoplasm, the rod is thought to grow to great lengths while at the same time branching in an apparently unlimited and uncontrolled fashion (Smith *et al.*, 1964).

Because the above study was carried out on purified suspensions of the long virus threads and their concomitant capsules, the criticism could be made that some at least of the above phenomena were artifacts and the elaborate branching of the long threads might be due to aggregation of different lengths. The investigation has now been repeated, but this time on sections of infected cells because if the phenomena described above take place *inside* the cell they cannot be ascribed to

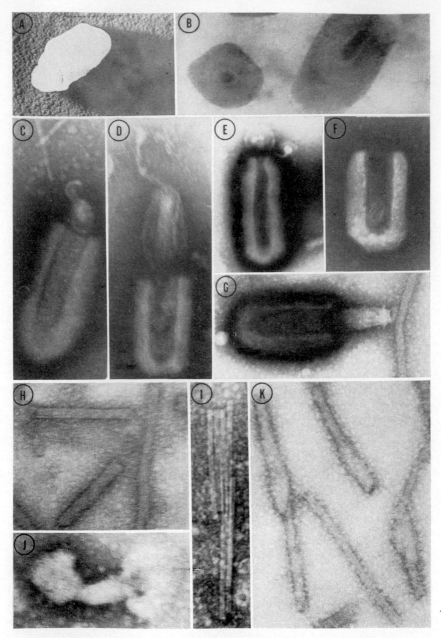

FIG. 36. See legend opposite page.

adsorption or other artifacts. The sequence of events postulated from an examination of the purified preparations has now been observed in thin sections of granulosis-infected fat-body cells of several species of larvae including *Pieris rapae* and cutworms (Noctuidae).

In sections of infected fat-body cells numerous examples have been observed of the same apparent extrusion of the rod as was seen in purified suspensions. As the rod emerges from the crystal it presumably increases in length (Figs. 37, 38). In the cell cytoplasm it undergoes unlimited and uncontrolled branching leading to the production of bizarre formations (Figs. 39, 40). Negative staining with PTA seems likely to rupture the long threads, revealing what appears to be an internal helix.

There are various possible explanations for the function of the long virus rods; they may represent an alternative mode of replication, in which the branching is a method of increasing the amount of DNA. Another possible explanation is that the virus exists in two forms, somewhat after the fashion of influenza virus.

Another theory which does take into account the curious branching is one suggested by Prof. F. Pautard, that here we have a kind of "molecular sickness." The remarkable branching of the virus threads may indicate an irregular disturbance of the replication mechanism, causing dislocations to occur in the core or causing overproduction of protein. If the branching were such as to produce a pattern of molecular regularity, then this might be a normal arrangement generated by a regular folding in the macromolecules. The apparently random character of the branching, however, suggests a sporadic failure of an otherwise

---

FIG. 36. Granulosis Virus. A. The intact capsule shadowed with uranium. Magnification, × ca 75,000. B. Sections of two capsules showing the occluded virus rod surrounded by the outer membrane. Magnification, × 118,000. C. Capsule with the cap removed showing the curious "appendage" at the end of the virus rod. Magnification, × 118,000. D. Similar to C, but the virus rod, together with the outer membrance, is now halfway out of the capsule; note that the "appendage" seems to have uncoiled. Magnification, × 111,000. E. Capsule showing the cap and "appendage." Magnification, × ca 68,200. F. Empty capsule. Magnification, × 118,000. G. Another aspect of the emergence of the virus rod from the capsule. Magnification, × 115,000. H. The outer membrane and two short virus rods. Magnification, × 111,000. I. Several thin rods lying side by side. Magnification, × 127,-000. J. The "appendage" and part of the cap at higher magnification. Magnification, × 180,000. K. Some of the long rods showing the "swellings." Magnification, × 118,000. (From Smith, Trontl, and Frist, 1964.)

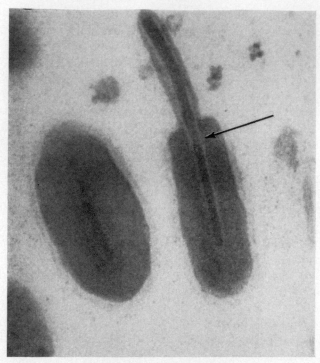

FIG. 37. Sections of two granulosis capsules side by side in the same cell: note the swelling at the end of the left-hand capsule which may be part of the removable cap. In the second capsule the rod is apparently emerging; note that the outer membrane now covers only half the rod (arrow). Magnification, × 110,000. (From Smith and Brown, 1965.)

linear mechanism, and this might be called a disease, or at least, the first signs of molecular sickness.

This possibility is the only one which attempts to give a reason for the uncontrolled and unlimited branching which leads to fantastic formations of no apparent practical use (Smith and Brown, 1965b).

D. Noninclusion-Body Viruses

We have mentioned earlier that with the development of an insect-tissue culture system we should gain an insight into the methods of insect virus replication. We shall have occasion to refer later in this section to the growth of *Sericesthis* iridescent virus in tissue culture.

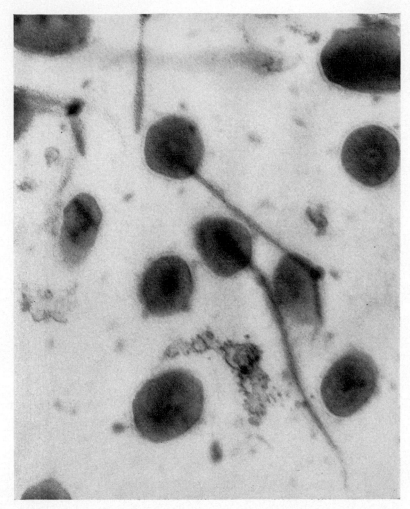

FIG. 38. Sections through two granulosis capsules in an infected cell, showing the apparent elongation of the rod as it emerges from the capsule. Magnification, × 46,-000. (From Smith and Brown, 1965b.)

a. *Tipula* IRIDESCENT VIRUS (TIV)

There is one phenomenon which has been observed in common in many spherical or near-spherical viruses from both plants and animals, and that is the appearance of numbers of empty protein shells of the

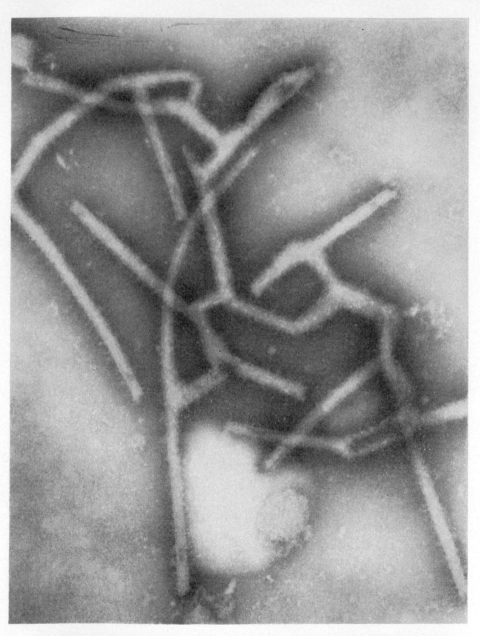

FIG. 39. A long virus rod showing the bizarre shapes produced. Note the angle of the branching and that part of the rod is ruptured. Magnification, × 50,000. (From Smith and Brown, 1965b.)

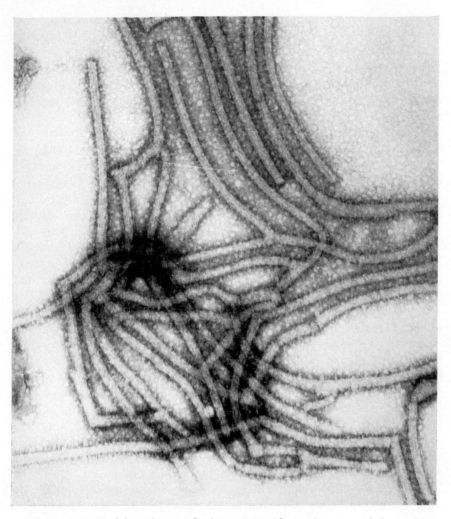

FIG 40. A mass of branching rods from a granulosis of a noctuid larva. Magnification, × 80,000. (From Smith, Trontl, and Frist, 1964.)

same size as the virus. This phenomenon was first observed by Markham and Smith (1949) with the plant virus causing turnip yellow mosaic. These empty shells, called "top component" because they form the top layer during ultracentrifugation, do not contain RNA and are therefore not infectious. Somewhat similar empty shells have been described by Horne and Nagington (1959) in studies on the structure and development of poliovirus. The highest count of empty shells they recorded was 11–14% at 5½ hours after infection.

Examination of ultrathin sections of fat-body cells from *Tipula paludosa* in an early stage of infection with TIV reveals the presence of empty membranes in large numbers; they are often more numerous than the normal virus particles.

Careful study of the empty forms of TIV suggests very strongly that they are stages in the reproduction of the virus. This is borne out by the following facts. (1) Empty or partially empty shells occur predominantly in the early stages of the disease. (2) What appear to be stages in the development of the contents of the particle can be seen (Fig. 41); in ultrathin sections, intermediate forms with the central core varying greatly in size were found (Smith and Hills, 1962).

Xeros (1964) has also studied the development of TIV particles and agrees that the empty membranes are formed first. He suggests, however, that the empty membranes are not complete when first formed, but have an opening into which the DNA-protein is introduced. The membrane is thus not completed until the full complement of nucleoprotein is occluded.

Massive virogenic stromata are formed, somewhat similar to those described in the other types of insect virus diseases, but the growth of the stroma and the enlargement of the cell are unusually great. So, of course, is the quantity of viruses produced from the stroma of each cell. As the stromata mature the formation of the viral membranes is apparent; they appear as cup-shaped membranes, somewhat wider than mature virus particles, within the dense stromal masses. As the virus particles form, the coherent fibrillar stroma disappears, the fibrils apparently becoming tightly coiled within the viral membranes. In addition chains of elongated microvesicles have been observed traversing the mature stroma. These are found in the stroma before the development of either virus particles or viral membranes. It is from these microvesicles that the viral membranes are thought to develop (Xeros, 1964).

b. *Sericesthis* IRIDESCENT VIRUS (SIV)

The development of SIV in the blood cells of *Galleria mellonella* L. has been studied by the electron microscopy of thin sections (Leutenegger, 1964). The particles develop in the cytoplasm of the blood cells, and similar developmental stages to those described for TIV (Smith and

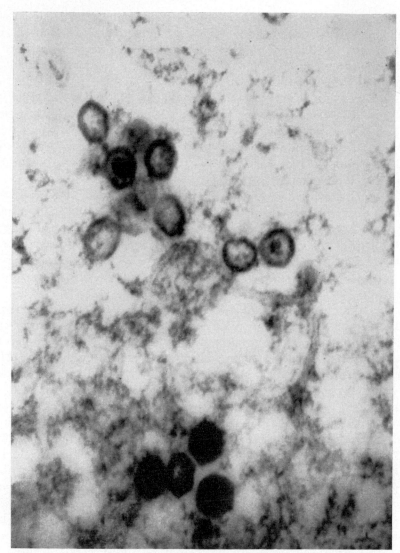

Fig. 41. Electron micrograph of sections of the *Tipula* iridescent virus particles showing apparent developmental stages. Magnification, × 80,000.

Hills, 1962) can be demonstrated. Six hours after inoculation a few immature viral particles occur in the cytoplasm. Forty-eight hours after infection the number of virus particles has increased and mature particles along with various immature forms are present. The infection leads to striking cytoplasmic changes such as the formation of dense areas and membranous structures. The dense granulated areas seem to be sites of virus replication and are probably the same as the virogenic stromata in which the particles of TIV are formed.

The multiplication of SIV in cell cultures of *Antheraea eucalypti* Scott has been investigated by Bellett and Mercer (1964). This is almost the first occasion on which the development of an insect virus has been followed in tissue culture. It is to be hoped that soon there will be developed a tissue-culture system similar to that in everyday use for the viruses of higher animals.

Cultures of ovarian cells derived by Grace from *A. eucalypti* were successfully infected with SIV and studied by means of fluorescent antibody and acridine orange staining and by electron microscopy.

Discrete foci of virus antigen and DNA developed in the cytoplasm of infected cells about 2 days after infection at 25°C or 3 days at 21°C; the proportion of cells containing foci increased to reach a maximum after 4 days of incubation at 25°C or 6 days at 21°C.

The size and intensity of fluorescence of the foci increased until the cytoplasm of many cells was filled with viral products. The cytoplasmic foci were shown by electron microscopy to contain aggregations of complete virus particles, developmental forms resembling those of TIV, and diffuse material which stained with uranyl acetate. Viral particles, apparently in several stages of assembly, were seen in the dense areas; these ranged from almost empty shells through partly filled to completely filled shells. The closing shell appeared to segregate a sample of the dense material in its neighborhood. "Cartwheel" forms similar to those described by Smith and Hills (1962) were also observed. The replication process of SIV seems therefore very similar to that described for TIV by Smith (1958a,b), Smith and Hills (1962), and Xeros (1964). A phenomenon observed in tissue culture, but which has not been described in the infected insect, was the presence of single virus particles in long processes extending from the cell surface. They appeared to separate from these by a budding process and acquired an envelope derived from the cell membrane.

Seven consecutive passages of SIV in *A. eucalypti* cell cultures were successful at 21°C, but infectivity was lost after two passages at 25°C.

To study multiplication of SIV *in vitro* it is necessary to develop a quantitative assay for infective virus in the cell-culture system. Since plaque titration of SIV has not been achieved and the cytopathic effect is not always accompanied by virus multiplication, assay must be based on the proportion of cells stained by fluorescent antibody.

The optimum temperature for production of the antigen of SIV by infected cell cultures is 20°C. Virus titers in infective units (IU) per milliliter can be estimated from the proportion of cells which stain with fluorescent antibody 6½ days after inoculation with dilutions of the virus (Bellett, 1965a).

In quantitative experiments with SIV in tissue culture, Bellett (1965b) found that viral DNA was synthesized in cytoplasmic foci which also contained antigen. Each infective unit (IU) of virus formed a separate focus, with supernumerary foci in some cells. The initiation of viral replication was a random process, and once replication was initiated in a cell with multiple infection, all the foci began to operate at about the same time. Synthesis of viral DNA and antigen was detected in some cells 2 days after inoculation; most infected cells had begun viral replication by 6 days and continued to synthesize viral DNA for a further 2 days. The host-cell nuclear DNA was not inhibited during the synthesis of the viral DNA.

There is an eclipse period of about 4 days after inoculation of the cell culture; infective virus is then continuously produced and released from cells until the eighth day. The yield of virus is about 500 IU per infected cell, of which about 360 IU per cell are released.

REFERENCES

Aizawa, K. (1953). Multiplication mode of the silkworm jaundice virus. I. On the multiplication mode in connection with the latent period and the LD_{50}-time curve. *Sanshi Shikensho Hokoku* 14, 201–228. (English summary).

Aizawa, K. (1959). Mode of multiplication of silkworm polyhedrosis virus. II. Multiplication of the virus in the early period of the LD_{50}-time curve. *J. Insect Pathol.* 1, 67–74.

Ayuzawa, C. (1961). On the resistance to infection with nuclear polyhedrosis in the silkworm *Bombyx mori* L. *Nippon Sanshigaku Zasshi* 30, 413–419. (English summary).

Bellett, A. J. D. (1965a). The multiplication of *Sericesthis* iridescent virus in cell cultures from *Antheraea eucalypti* Scott. II. An *in vitro* assay for the virus. *Virology* 26, 127–131.

Bellett, A. J. D. (1965b). The multiplication of *Sericesthis* iridescent virus in cell cultures of *Antheraea eucalypti* Scott. III. Quantitative experiments. *Virology* 26, 132–141.

Bellett, A. J. Ď., and Mercer, E. H. (1964). The multiplication of *Serciesthis* iridescent virus in cell cultures from *Antheraea eucalypti* Scott. I. Qualitative experiments. *Virology* 24, 645–653.

Bergold, G. H. (1950). The multiplication of insect viruses as organisms. *Can. J. Res.* E28, 5–11.

Bergold, G. H. (1958). Viruses of insects. *In* "Handbuch der Virusforschung" (C. F. Hallauer and K. F. Meyer, eds.), Vol. 4, pp. 60–142. Springer, Vienna.

Bird, F. T. (1959). Polyhedrosis and granulosis viruses causing single and double infections in the spruce budworm. *J. Insect Pathol.* 1, 406–430.

Bird, F. T. (1964). On the development of insect polyhedrosis and granulosis virus particles. *Can. J. Microbiol.* 10, 49–52.

Franklin, R. E., Caspar, D. L. D., and Klug, A. (1959). *Plant Pathol., Probl. Progr., 1908–1958* p. 447, University Press, Wisconsin.

Horne, R. W., and Nagington, J. (1959). Electron microscope studies of the development and structure of poliomyelitis virus. *J. Mol. Biol.* 1, 333.

Krieg, A. (1955). Untersuchungen uber die Polyhedrose von *Neodiprion sertifer* (Geoffr.). *Arch. Ges. Virusforsch.* 6, 163–174.

Krieg, A. (1958). Latente und akute Infectionen mit Insekten-Viren. *Proc. 10th Intern. Congr. Entomol., Montreal, 1956* Vol. 4, pp. 737–740. Intern. Congr. Entomol., Ottawa, Canada.

Krieg, A. (1961). Uber den Aufbau und Vermehrungsmoglichkeiten von stabchenformigen Insekten-Viren. *Z. Naturforsch.* 12b, 120–121.

Leutenegger, R. (1964). Development of an icosahedral virus in hemocytes of *Galleria mellonella* (L.). *Virology* 24, 200–204.

Markham, R., and Smith, K. M. (1949). Studies on the virus of turnip yellow mosaic. *Parasitology* 39, 330.

Paillot, A. (1930). "Traité des maladies du ver a soie," 279 pp. Doin, Paris.

Pautard, F. (1965). Personal communication.

Smith, K. M. (1958a). Early stages of infection with the *Tipula* iridescent virus. *Nature* 181, 996–997.

Smith, K. M. (1958b). A study of the early stages of infection with the *Tipula* iridescent virus. *Parasitology* 48, 459–462.

Smith, K. M., and Brown, R. M. (1965a). Replication cycle of an insect granulosis virus. *Nature* 207, 106–107.

Smith, K. M., and Brown, R. M. (1965b). A study of the long rods associated with insect granuloses. *Virology* 27, 512–519.

Smith, K. M., and Hills, G. J. (1962). Replication and ultrastructure of insect viruses. *Proc. 11th Intern. Congr. Entomol., Vienna, 1960* Vol. 2, pp. 823–827.

Smith, K. M., and Xeros, N. (1953). Development of virus in the cell nucleus. *Nature* 172, 670–671.

Smith, K. M., and Xeros, N. (1954). Electron and light microscope studies of the development of the virus rods of insect polyhedroses. *Parasitology* 44, 71–80.

Smith, K. M., Wyckoff, R. W. G., and Xeros, N. (1953). Polyhedral virus diseases affecting the larvae of the privet hawk moth (*Sphinx ligustri*). *Parasitology* 42, 287–289.

Smith, K. M., Trontl, Z. M., and Frist, R. H. (1964). A note on a granulosis virus disease of a noctuid larva. *Virology* **24**, 508–513.

Xeros, N. (1955). Origin of the virus-producing mass or net of the insect nuclear polyhedroses. *Nature* **175**, 588.

Xeros, N. (1956). The virogenic stroma in nuclear and cytoplasmic polyhedroses. *Nature* **178**, 412–413.

Xeros, N. (1964). Development of the *Tipula* iridescent virus. *J. Insect Pathol.* **6**, 261–283.

Yamafuji, K., Yoshihara, F., and Sato, M. (1954). Eclipse period of polyhedral disease. *Enzymologia* **17**, 152–154.

CHAPTER VII

Transmission and Spread of Insect Viruses

A. Infection *per os*

Ingestion of contaminated food or other material is probably one of the most effective means of virus transmission in nature, at all events so far as the inclusion-body viruses are concerned.

There seems to be some confusion of thought regarding the spread of nuclear polyhedroses by ingestion. According to some workers virus infectivity has not been demonstrated in the feces of virus-diseased larvae such as those of *Lymantria monacha* and *Bombyx mori* (Escherich and Miyajima, 1911; Aizawa, 1953a). This is said to be due to an antiviral substance in the gut juice of the silkworm (Aizawa, 1953b, 1962a; Masera, 1954).

On the other hand, polyhedra cause infection when administered *per os*. The polyhedra are supposedly dissolved by the alkaline gut juice, and the liberated virus particles penetrate through the gut epithelium and multiply in the cells of the blood and other tissues (Day *et al.*, 1958; Vago and Croissant, 1959; Bird, 1959).

It is difficult to reconcile the statement that the gut juice inactivates the virus with the other view that the gut juice dissolves the polyhedra and liberates the virus particles to penetrate the gut epithelium. There is, however, no doubt that ingestion of the polyhedra, both nuclear and cytoplasmic, and of the capsules of granuloses gives a high rate of infection.

In a study of the transmission of nuclear polyhedrosis virus in laboratory populations of *Trichoplusia ni* (Hübner), Jaques (1962) found that the feces of the larvae did contain virus and the infectivity of

excrement from infected larvae was dependent on the dosage of polyhedra ingested but not on the period between ingestion of polyhedra and defecation. An additional 100-fold increase in numbers of
polyhedra ingested substantially increased the infectivity of the excrement.

The proportion of the leaf surface covered with polyhedra, as well
as the location and distribution of the deposits, has considerable effect on
mortality. As the area of centrally located deposits is increased from
1 to 10 cm², mortality increases threefold. Location of deposits of polyhedra at the periphery of leaf disks rather than in the center resulted
in higher mortality (Jacques, 1962). This would be expected since
caterpillars usually start to feed at the leaf edge.

In the nuclear polyhedroses and the granuloses which liquefy the
body contents a lot of virus ingestion must result from the contamination
of the food by the disintegrated bodies. Furthermore, in the granuloses, the liquefied cadavers seem to have an attraction for the larvae
of *Pieris brassicae*, the large white butterfly, which feed greedily on
the remains (Smith, 1959, 1960; Wilson, 1960).

Whatever the situation may be about the infectivity of the feces in
nuclear polyhedroses, there is no doubt concerning the presence of cytoplasmic polyhedra in the excrement of infected larvae. In a cytoplasmic
polyhedrosis the polyhedra are situated in the cells of the midgut and
as the disease develops the polyhedra are liberated into the gut lumen
and voided with the feces. Furthermore, in the late stages of the disease
the gut is so loaded with polyhedra that large numbers are regurgitated
by the larvae.

In the case of the *Tipula* iridescent virus (TIV), a noninclusion-
body virus, transmission *per os* is much less efficient than injection, and
feeding caterpillars of *P. brassicae* with contaminated foliage gave only
a low rate of infection. However, this could be increased in the following manner. When disturbed, these larvae regurgitate a drop of blood
which remains for a short period suspended from the mouth. If this
drop is withdrawn by means of a pipette and a drop of purified virus
rapidly substituted, the virus is taken in by the caterpillar in the place
of the drop of regurgitated blood (Smith *et al.*, 1961).

The two viruses affecting adult honeybees, acute bee paralysis virus
(ABPV) and chronic bee paralysis virus (CBPV) are easily transmitted
to the bees by feeding them syrup contaminated with virus (Bailey
et al., 1963).

Similarly, the larvae of honeybees can be infected with sacbrood virus (SBV) by putting 1 μl of virus suspension into the food surrounding larvae (Bailey *et al.*, 1963). White (1917), and subsequently others, caused sacbrood in larvae of healthy bee colonies by feeding the adults with a mixture of syrup and bacteria-free extract of diseased larvae.

B. Transovarial Transmission

For a time transovarial transmission of insect viruses was a controversial subject, but there is now so much evidence that "vertical transmission," as it is called, does take place with a number of insect viruses that it is now generally accepted.

While the fact of transovarial transmission is now accepted, there is still some argument as to whether the virus is contained inside the egg or adsorbed to the egg surface. Probably both are true; the term "transovarial" is used for transmission inside the egg and "transovum" for transmission by adsorption of the virus to the exterior of the egg. Transovarial dissemination of virus was first suggested as occurring in the silkworm, *B. mori*, by Conte (1907) and by Bolle (1908).

A good deal of investigation has been carried out on the generation-to-generation spread of the cytoplasmic polyhedrosis virus of *B. mori*, the silkworm. Hukuhara (1962) inoculated fifth-instar larvae with virus in either hexagonal or tetragonal cytoplasmic polyhedra. Since it is thought that the virus dictates the shape of the polyhedra, and not the cell, the type of polyhedra can be used as a marker. This helps to d:stinguish the inoculated virus from possible latent virus infections. Some of the inoculated larvae died but some completed their metamorphosis and emerged as adults. In many cases polyhedra could be observed in the midgut epithelium of these adults.

In the progeny of silkworms which were inoculated with hexagonal cytoplasmic polyhedra, 23% of the test larvae developed cytoplasmic polyhedrosis. Seventy-five per cent of the infected larvae contained only hexagonal polyhedra in the midgut and the remaining 25% contained both tetragonal and hexagonal polyhedra.

On the other hand, in the progeny of silkworms which were inoculated with tetragonal polyhedra, some of the infected larvae contained only tetragonal polyhedra, though most of the infected larvae contained only hexagonal polyhedra or both hexagonal and tetragonal polyhedra.

On the basis of these results Hukuhara thinks that normal silkworms transmit hexagonal cytoplasmic polyhedron virus to the next generation in an occult state through the germ cell, and that when normal larvae are inoculated with the tetragonal polyhedron virus, some of the larvae transmit the tetragonal polyhedron virus to a part of the next generation in an occult state.

In a somewhat similar experiment, larvae of *B. mori* which had been infected in the first instar with cytoplasmic polyhedrosis virus were able to transmit the virus to the next generation. After cold treatment (5°C) for 24 hours, the occurrence of cytoplasmic polyhedrosis (tetragonal polyhedron virus) was observed in larvae of the fifth instar of the F_1 generation.

The frequency of cytoplasmic polyhedrosis in the F_1 generation was markedly higher in the case where females from inoculated stock (infected *per os*) were used than in the controls in which no virus had been inoculated in the previous generation. The shape of the polyhedra observed in the F_1 generation had the same tetragonal outline as that of the polyhedra inoculated to the previous generation.

These results suggest that the cytoplasmic polyhedrosis viruses in *B. mori* are transmitted to the next generation through the egg, apparently in an occult state (Aruga and Nagashima, 1962). Tanada *et al.* (1964) carried out a survey of the presence of a cytoplasmic polyhedrosis virus in field populations of the alfalfa caterpillar, *Colias eurytheme* Boisd. They found that certain individuals, among adults collected from 12 localities covering nearly the entire length of California, were found to be carriers of a latent cytoplasmic polyhedrosis virus. The collections represented five populations which lived in more or less isolated valleys. The virus apparently occurred in these populations either in an occult or chronic condition.

It has been shown (Smirnoff, 1962b) that the later instars of the larvae of the sawfly, *Neodiprion swainei*, do not die after infection with weak concentrations of a nuclear polyhedrosis virus and about 60% of the treated larvae develop into normal adults. These retain the ability to transmit the virus through the egg. Infection in larvae hatched from eggs laid by female carriers of the virus develops after approximately 13 days and is about 90% fatal.

Some further experiments on the generation-to-generation transmission of a nuclear polyhedrosis virus have been carried out by Harpaz and Ben Shaked (1964). The virus in question attacks the larvae of *Prodenia litura* Fabr., and was first described by Smith and Rivers

(1956). Egg masses laid by two females of different origin were each divided into two. One half was surface sterilized with KOH and reared aseptically on lucerne meal medium; the other half was reared on fresh cotton leaves. The larvae were reared in groups of 25, each larva being kept in a separate container to avoid contagion. In the results no significant difference was found in the amount of polyhedrosis whether the larvae were reared aseptically or not. There was, however, a highly significant difference in the two genetic lines 1 and 2. In larvae reared on an aseptic diet the number of cases of polyhedrosis was 3 times greater in line 2 than in line 1, and in larvae raised on a cotton leaf diet the number of infections in line 2 was 5 times that of line 1. In view of the possibility that the surface sterilization of the eggs was insufficient to sterilize any possible polyhedra adhering to the chorion, a further experiment was performed. Groups of 25 third-instar larvae were fed various homogenates. These consisted of whole-egg homogenates from virus-infected stock, whole-egg homogenates from virus-free stock, and egg-shell homogenates from virus-infected stock. The results indicated that none of the three types of egg-matter homogenates could provoke any significantly higher frequency of polyhedrosis than the natural spontaneous frequency of the untreated control series. The results of these experiments indicate that generation-to-generation transmission of this virus is not merely a mechanical contamination of the egg surface but that a more complicated genetic mechanism is involved.

There are other well-documented instances of transovarial dissemination. Larvae of *Telea polyphemus* Cram. have been observed to die of a nuclear polyhedrosis within 48 hours of emergence from the egg (Smith and Wyckoff, 1951). In some cases, in which the larvae have failed to emerge from the egg, examination revealed the presence of large numbers of polyhedra (Rivers, 1962).

It is reasonable to suppose that if the polyhedra were on the external surface of the egg and were ingested by the emerging larvae there would be an incubation period of 5–10 days before the development of the disease, but this was not the case with the instances cited above. There are other cases where surface sterilization of the egg by NaOH, HCl, formaldehyde, and trichloracetic acid has failed to produce virus-free progeny. Rivers reports experiments with a granulosis and the eggs of *Pieris brassicae* Linn., in which the chorion was entirely removed so that the embryo was exposed; in spite of this a proportion of the resulting larvae were infected with the granulosis virus. In an experiment carried out by C. F. Rivers and the writer on the use of a nuclear polyhedrosis to control the larvae of the sawfly *Neodiprion sertifer* (Geoffroy) on pine trees, the virus was introduced into an area where it had not previously existed. At the end of the season a few female sawflies were collected and brought back to the laboratory where the eggs were collected and placed under virus-free conditions. A high proportion of the larvae resulting from these eggs died of a nuclear polyhedrosis.

The situation regarding the transmission through the egg of the non-inclusion viruses is less clear-cut. Bailey (1965) has managed to establish chronic bee paralysis virus (CBPV) in bee colonies by substituting a queen from colonies found elsewhere with the disease. This is interesting confirmation that the disease is inherited.

The *Tipula* iridescent virus (TIV) has been observed in the head, near the eyes, of adult white butterflies, *Pieris brassicae* (Fig. 42) but there is no evidence that this virus is transmitted through the egg. Indeed, investigation by Day (1965) on the very similar *Sericesthis* iridescent virus (SIV) suggests that the virus is not transmitted through the egg, at all events in one species. Adult individuals of *Galleria mellonella,* the greater wax moth, in which all epithelial cells were infected with SIV, survived normally, mated, and oviposited. The larvae hatched normally, and attempts were made to isolate virus from them without success. Thus SIV remains infectious after the metamorphosis of the host, but transovarial infection was not observed. Day points out that in view of the absence of viral DNA from the reproductive tissues, as revealed by Feulgen staining this is not surprising. It can therefore be concluded that transovarial infection of SIV does not occur in *G. mellonella.*

C. Artificial Methods of Transmission

Burnside (1945) demonstrated that the virus of bee paralysis could be transmitted by spraying a bacterium-free extract of paralyzed bees onto adult bees in the hive. Since then Bailey *et al.* (1963) have used this method to infect bees with both paralysis viruses, acute bee paralysis virus (ABPV) and chronic bee paralysis virus (CBPV). It is not clear exactly how the sprayed virus enters the bees; infection by way of the spiracles seems unlikely and contamination of the mouthparts may be the route of infection.

By far the most successful method of artificial transmission is by injection. The preparation of suitable needles for this purpose has been described by Maramorosch (1951) and by Martignoni (1955). These consist essentially of glass capillary tubes cemented to hypodermic needles which may be either intact or cut off 6 mm from their bases. Maramorosch used a paste of glycerine and litharge for cementing the glass and metal together but Martignoni found sealing wax adequate.

In another paper Martignoni (1959) suggests a more elaborate needle for microinjection, one which will stand heat sterilization. There are two procedures which are described in detail in his paper.

Very good results, however, can be achieved by the use of the finest steel hypodermic needles, about 0.01 or 0.02 ml of virus suspension being injected at a time. Success depends a good deal on the purity and concentration of the virus inoculum. The best results, in the case of the noninclusion-body viruses, are obtained after centrifuging the virus on a sucrose gradient; the death rate due to extraneous bacterial infection can be greatly reduced by the addition to the inoculum of 200 IU of penicillin and 200 IU of streptomycin per milliliter, respectively.

In using this method of infection with the inclusion-body viruses, it seems that injection of nuclear polyhedra into the body cavity does not cause disease, but subcutaneous injection does.

It seems that infection with granulosis virus does not occur when injections of the capsules are made into the larval hemocoele but it does with peroral application (Martignoni, 1957).

According to Bergold (1958), quoted by Aizawa (1963), the LD_{50} in infection with polyhedra is 50 to 100 μg (2.5×10^{-6} gm of virus) in *Bombyx mori* and 10 μg in *Porthetria dispar* and *Malacosoma disstria* (Hübner). In *Neodiprion sertifer* the LD_{50} is 100–500 polyhedra, according to Bird and Whalen (1953) and 50–100 polyhedra according to Krieg (1955).

The amount of the *Tipula* iridescent virus which must be injected to give infection varies according to the species of insect under consideration. It was found that some species of insect required higher virus concentrations than others to become infected. For example, the larvae of *Sphinx ligustri* Linn. were easily infected with a dose of 8×10^9 virus particles but not with one of 8×10^6 particles which gave 100% infection with the larvae of *P. brassicae*. To infect the larvae of the silkworm *Bombyx mori* Linn. a dose of TIV similar to that used for *S. ligustri* was given (Smith *et al.*, 1961).

D. Cross Transmission

For a number of years the opinion has been strongly held by some insect virologists that these viruses are species-specific and that cross transmission does not occur. In view, however, of the mass of evidence to the contrary this view is no longer tenable.

The chief difficulty in resolving the problem of cross transmission lies in the lack of a reliable means of virus identification. In other words, is the virus which causes the death of an inoculated insect the same virus as the one injected? This subject is dealt with further in the next chapter concerning latency.

The obvious method of differentiating insect viruses is by means of serological relationships, but this has not yet been sufficiently developed although a start has been made (see Chapter X). However, there are other methods available, especially for the inclusion-body viruses. One of these is the fact, pointed out by Gershenson (1960) and others, that the shape of the polyhedral crystal is a function of the virus and not of the host cell. By comparing the shape of the polyhedra in the source insect with those produced in the inoculated insect, it would then be possible to say, with fair accuracy, whether a genuine cross transmission had taken place. Identification of the exact crystallographic shape of the polyhedra can be aided by a special staining technique which picks out the edges of the crystals (Gershenson, 1960).

Another method is to use a virus which has such distinctive qualities that it cannot be mistaken in whatever host insect it occurs. The *Tipula* iridescent virus (TIV) is particularly suitable for this purpose since, as we have already seen, it has unmistakable characteristics.

It is thus important when attempting cross transmission of inclusion-body viruses to choose a polyhedrosis virus with strongly characteristic polyhedra. It sometimes happens that in a given disease a few aberrant polyhedra will appear. These are thought to be due to a virus mutation, and when they are inoculated into fresh insects of the same species the polyhedra produced are all of the aberrant shape (Gershenson, 1960; Aruga *et al.*, 1961). This type of virus is especially suitable for cross transmission work.

So far as the nuclear polyhedroses are concerned, Gershenson (1959) has suggested that there may exist a group specificity where one virus is transmissible among members of the same genus. It certainly seems to be the case that the nuclear polyhedrosis of *Aglais urticae* Linn.), the small tortoiseshell butterfly, is intertransmissible among the vanessid butterflies, including *Vanessa atalanta* (Linn.), *V. cardui* (Linn.), and *Nymphalis io* (Linn.), the polyhedra in each species being identical. A few other examples of cross transmission with nuclear polyhedrosis viruses follow. Stairs (1964) has carried out infection experiments with a nuclear polyhedrosis affecting species of tent caterpillars *Malacosoma* spp. Although cross infection appears to be established in the genus *Malacosoma*, the virus from one species may vary in its pathogenicity towards another. Nuclear polyhedrosis viruses from *Malacosoma americanum* (Fabricius), *M. pluviale* Dyar, and *M. disstria* (Hübner) were equally infectious for *M. disstria*. The virus from a European species *M. alpica* Staudinger, was infectious for *M. disstria* but acted much more slowly than any of the native viruses.

Larvae of the clouded drab moth *Orthosia incerta* (Hufnagel) (Noctuidae) were found to be susceptible to a nuclear polyhedrosis virus from *Barathra brassicae* (Linn.) (Ponsen and deJong, 1964).

Similarly a nuclear polyhedrosis virus from *Erannis tiliaria* (Harris) was transmissible to two other species of Geometridae, *Alsophila pometaria* (Harris) and *Phigalia titea* Cramer. The shape and dimensions of the polyhedra in these two geometrids were similar to those found in the source insect *Erannis tiliaria*. Attempts to transmit the virus to six other species of lepidoptera were not successful (Smirnoff, 1962a, 1964). The same author (Smirnoff, 1963) reports the transmission of a nuclear polyhedrosis virus from the larvae of the poplar sawfly, *Trichiocampus viminalis* Fall., to the larvae of the willow sawfly *Trichiocampus irregularis* (Dyar).

Steinhaus (1953) reports the susceptibility of the alfalfa larva *Colias philodice eurytheme* Boisd. to a nuclear polyhedrosis virus from the larva of a South American species of the same genus, *Colias lesbia* Fabr.

The larvae of the greater wax moth *Galleria mellonella* Linn., are susceptible to infection with a nuclear polyhedrosis virus from the silkworm, *B. mori*, and the virus titer was determined during the course of serial passages in *Galleria*. This virus was infectious for both *Bombyx* and *Galleria*, and it was neutralized by an antiserum against silkworm nuclear polyhedrosis virus but not by an antiserum against silkworm cytoplasmic polyhedrosis virus (Aizawa, 1962b).

Cross transmission, however, is not confined to related groups of insects in the same families and genera; and, if we accept the criterion of similar polyhedra indicating similar viruses, a most striking example of the cross transmission of a nuclear polyhedrosis virus is the infection of larvae of *Hemerobius* spp. (Neuroptera) from larvae of *Porthetria dispar* Linn. (Lepidoptera). In this case not only were the polyhedra identical, but also the arrangement of the bundles of virus rods inside the polyhedra (Sidor, 1960; Smith *et al.*, 1959).

The situation with regard to the cross transmission of the cytoplasmic polyhedrosis viruses is intimately bound up with the question as to how far many of these viruses are identical. It is certainly true that cross transmission seems to take place between various species of tiger moth larvae (Arctiidae), but final judgment in these cases must wait upon the development of criteria of relationship, either by serological or chemical data.

It may be relevant here to give the results of one or two experiments out of a large number performed in 1961 by the writer on attempted cross transmission of cytoplasmic polyhedrosis viruses. In the first experiment the insect used was the comma butterfly, *Polygonia c-album* Linn., and all the larvae were the progeny of a single female. On April 6, 30 young (second-stage) larvae were colonized on nettle leaves contaminated with cytoplasmic polyhedra from *Arctia caja*, the tiger moth. Between April 17 and April 28 6 larvae died of a cytoplasmic polyhedrosis, and on May 11 a butterfly emerged and died shortly afterward; the gut of this butterfly was found to be filled with cytoplasmic polyhedra. The remaining larvae pupated normally.

In the second part of this experiment, 30 similar larvae were fed with cytoplasmic polyhedra from *Phalera bucephala*, the buff-tip moth. On April 17, 11 days after the start of the experiment, 7 larvae were dead of a cytoplasmic polyhedrosis and the rest were very much smaller and more backward than the controls, all of which appeared healthy, some having already pupated. By April 21 all the larva which had been fed with the cytoplasmic polyhedra from *P. bucephala* were dead of a cytoplasmic polyhedrosis.

Of the control larvae all but two pupated normally; these two died of a cytoplasmic polyhedrosis. This experiment was repeated using the polyhedra which developed after the larvae fed on the polyhedra from *P. bucephala*. In this experiment 13 larvae out of 24 died of a cytoplasmic polyhedrosis; two of the controls were similarly affected.

In another experiment 35 second-stage larvae of *Polygonia c-album* were fed with cytoplasmic polyhedra from *Sphinx ligustri*, the privet hawk moth; these polyhedra had developed in the *S. ligustri* larvae after feeding on nuclear polyhedra from a larva of *Nymphalis io*, the peacock butterfly. This experiment was commenced on May 15 and by May 25, 27 larvae out of the 35 had died of a cytoplasmic polyhedrosis. Of the 35 control larvae, 12 died of a cytoplasmic polyhedrosis.

Experiments with the larvae of another species, *Gonepteryx rhamni* Linn., the brimstone butterfly, were commenced on April 11, cytoplasmic polyhedra from *Arctia caja* being again used. By April 24, the experimental larvae showed a most impressive difference from the controls, being only about half the size. By April 28 all the experimental larvae had died of a cytoplasmic polyhedrosis. On April 25, the control larvae were nearly full-grown and appeared healthy; 28 of these pupated normally, but two died of a cytoplasmic polyhedrosis.

The obvious difficulty in assessing the results of these transmission experiments is the lack of means of differentiating between the viruses. Were these genuine cross transmissions or merely a stimulation of a latent virus infection? The mortality could not have been much higher if the larvae had been fed with a wild virus from their own species. It was known that both *Polygonia c-album* and *G. rhamni* do have cytoplasmic polyhedroses, and this is shown by the fact that some of the control larvae of the two species developed the disease spontaneously. Nevertheless the high rate of mortality does suggest that cross transmission had taken place and is further evidence that many of these cytoplasmic polyhedrosis viruses are one and the same.

Aruga *et al.* (1961) reported that they had transmitted the cytoplasmic polyhedrosis of the silkworm to *Antheraea pernyi* Guerin-Melville. They observed, however, that the shape of the polyhedra in *A. pernyi* was not uniform and differed from that of the silkworm polyhedra used for the inoculation. If we accept the theory that the shape of the polyhedra is a characteristic of the virus then the above experiment suggests an activation of an occult virus rather than a true cross transmission.

Tanada and Chang (1962) made cross-transmission studies with three cytoplasmic polyhedrosis viruses. They consider that a true cross transmission was made with two of the viruses, one from the alfalfa caterpillar (*Colias eurytheme* Boisd.) and the other from the armyworm (*Pseudaletia unipuncta* Haworth), for the following reasons.

1. The reciprocal transmission of the two viruses to the two insect hosts

2. The infectivity of the viruses to the original hosts after passage through the alternate hosts

3. The symptomatology and pathology in the two hosts, which are essentially characteristic for each host species regardless of whether the virus had been obtained from the original or alternate hosts

4. That the activation of a latent infection was unlikely was shown by test-feeding the larvae with normal midguts and a third virus, the cytoplasmic polyhedrosis virus of the silkworm.

Although the armyworm is susceptible to the cytoplasmic polyhedrosis virus of the alfalfa caterpillar, it is not infected by the cytoplasmic polyhedrosis of the silkworm.

In the discussion on cross transmission of nuclear polyhedrosis viruses an account was given of the apparent transmission of a virus from *Porthetria dispar* (Lepidoptera) to larvae of *Hemerobius* and *Chrysopa* spp. (Neuroptera). Similar cross-transmission experiments to neuropterous larvae, but using a cytoplasmic polyhedrosis from *Vanessa io,* the peacock butterfly, have been carried out (Smith, 1963). A peculiarity of the hemerobiid and *Chrysopa* larvae is that they lack an external opening to the gut during the larval life, and excretory matter is therefore accumulated in a large mass at the base of the hindgut. In larvae which had been fed on the cytoplasmic polyhedra, however, this mass had a different appearance from that of the control larvae. Sections of this mass, examined under the oil-immersion lens of the optical microscope, showed that it consisted entirely of polyhedra, representing a remarkable concentration of virus. Similar sections of control larvae revealed no polyhedral crystals. It seems, therefore, that in some cases, the cytoplasmic polyhedrosis viruses, like those of the nuclear polyhedroses, are transmissible not only between insect species but between different insect orders.

In 1956 a cytoplasmic polyhedrosis virus, originally taken from larvae of the painted lady butterfly, *Vanessa cardui* (Linn.) by the writer, was exported to Canada for testing against various forest insect pests. The virus was subsequently tried against 18 species of Lepidoptera and two of Hymenoptera; eleven of the Lepidoptera were found to be susceptible. Both cross transmission and back-cross transmission tests were made successfully even after the virus had been passed through two of the hosts at least 5 times. These transmission tests, along with the morphology of the polyhedra from different hosts and the histopathology in different hosts, suggest that true cross transmission of a nonspecific virus has been achieved rather than the induction of occult viruses (Neilson, 1964).

The situation as regards the intertransmissibility of the granulosis viruses is less clear. According to Huger (1963) they possess a high degree of host specificity. However, in the case of the pierids, *Pieris brassicae, P. rapae,* and *P. napi* one granulosis virus appeared equally infectious for all 3 species (Smith, 1959, 1960). In a study of the granulosis virus (Smith *et al.,* 1964) attempts to infect various species of cutworms (Noctuidae) with a granulosis virus from another species of cutworm were unsuccessful. On the other hand this same granulosis virus infected larvae of *Pieris rapae* with ease. It can, of course, always be said that this was not true transmission but the stimulation of a latent infection.

What is badly needed in a study of cross inoculation is a virus with definite characteristics which causes unmistakable symptoms; these would serve as a marker and differentiate the virus immediately from latent virus infections and casual contamination.

One of the noninclusion-body viruses is of this type; it is the *Tipula* iridescent virus (TIV) and as we have already seen in Chapter V it can be immediately recognized by the blue or violet iridescence of the affected organ, by the characteristic pentagonal or hexagonal outline of the virus particles in the electron microscope, and by the enormous quantities of virus produced in the diseased insect.

It was first shown with this virus that not only was cross transmission to different species achieved but also transmission between different orders of insects. In an investigation of the host range of TIV, the virus was transmitted experimentally to the larvae of 7 species of Diptera, 12 species of Lepidoptera, and 3 species of Coleoptera. The choice of experimental insects was governed largely by what insects were available, and there is little doubt that the experimental host range could be considerably widened (Smith *et al.,* 1961).

E. Methods of Spread

In this section we discuss briefly the methods of dissemination of the viruses in the field.

Probably the most efficient method of dispersal is the flight or migration of virus-carrying adults. We have mentioned already how females of the sawfly *Neodiprion sertifer,* when transported from the field into the laboratory, gave rise to diseased progeny. Tanada *et al.* (1964) demonstrated that females of the alfalfa butterfly *Colias eurytheme* were carriers of a latent infection with a cytoplasmic polyhedrosis virus which they spread among this species in the valleys in California.

Migratory movement of infected larvae is probably instrumental in spreading virus. The mass migrations of the armyworms and the movement of tent caterpillars from tree to tree are cases in point. Elmore and Howland (1964) studied the natural versus artificial dissemination of a nuclear polyhedrosis virus by contaminated adults of the cabbage looper. When the moths were contaminated externally in various ways with the virus a maximum of 51% of the progeny was diseased. When the moths were fed with virus-contaminated sugar water, 16 to 38% of their progeny were diseased. However, moths collected from a diseased field population did not transmit virus to their progeny. Dispersion of the virus by a foliage spray was more effective than dissemination by contaminated moths.

On the other hand when there is no possibility or special necessity of spraying, a virus may be introduced into an infesting colony by the insects themselves after they have been infected. In the case of a nuclear polyhedrosis of the sawfly *Neodiprion swainei* Middleton, Smirnoff (1962b) showed that once the virus had been introduced into the insect colonies, either by spraying or by the dissemination of infected cocoons, it was maintained from year to year by transovum transmission and was spread chiefly by the migration of older-instar larvae.

Martignoni and Milstead (1962) studied the transovum transmission of the nuclear polyhedrosis virus of *Colias eurytheme* Boisd. through artificial contamination of the female genitalia.

The rate of morbidity in the larvae resulting from the application of virus paste to the genital armature of adult females was determined on egg samples taken at 1 hour, at the second day, at the fourth day, and at the sixth day following application of the virus. This mortality, expressed in percentages of the emerged larvae, was 64.10 after 1 hour, 52.38 after 2 days, 21.95 after 4 days, and 19.51 after 6 days. Each female in the test group produced a mean of 56 larvae infected with nuclear polyhedrosis during a 6-day period.

The conclusion drawn from this experiment is that the spread of virus through the release of contaminated adults (as healthy carriers) is perhaps one of the most efficient ways for transmitting virus among members of a species.

A certain amount of virus spread is probably brought about by the action of parasites; this is almost equivalent to an injection since the virus is borne on the ovipositor of the parasite and is carried into the body of the larva during the process of egg-laying. As an example of this Thompson and Steinhaus (1950) showed that adults of the parasite *Apanteles* sp. could transmit virus to 3 successive larvae at hourly intervals. The polyhedra of certain nuclear polyhedroses can pass unchanged through the gut of a carnivorous bug, *Rhinoceris annulatus* Linn., and of the bird, *Erithacus rubecula* Linn. (Franz *et al.*, 1955). This, of course, helps the spread of the virus.

Weather conditions may also play a small part in virus dissemination, especially in the polyhedroses where the polyhedra may be spread around by the wind and the rain.

There is no doubt that the occlusion of the virus particles in the polyhedra materially aids both the dissemination and the longevity of the virus. The crystals protect the virus from the effects of an unfavorable environment such as inclement weather.

Steinhaus (1954, 1960) has shown that the polyhedra from a nuclear polyhedrosis of the silkworm retained infectivity for 20 years when stored at 4°C. When kept at room temperature for 37 years, however, infectivity was completely lost (Aizawa, 1954). Polyhedra of *Diprion hercyniae* (Hartig), a sawfly, show no infectivity after storage at 4.5°C for 12 years (Neilson and Elgee, 1960).

REFERENCES

Aizawa, K. (1953a). Silkworm jaundice virus in the excrements of the infected larvae. *Japan. J. Appl. Zool.* **18**, 143–144.

Aizawa, K. (1953b). On the inactivation of the silkworm jaundice virus. *Oyo Dobutsugaku Zasshi* **17**, 181–190. (English summary).

Aizawa, K. (1954). Dissolving curve and the virus activity of the polyhedral bodies of *Bombyx mori* L., obtained 37 years ago. *Sanshi-Kenkyu* **8**, 52–54. (English summary).

Aizawa, K. (1962a). Antiviral substance in the gut-juice of the silkworm, *Bombyx mori* (Linn). *J. Insect Pathol.* **4**, 72–76.

Aizawa, K. (1962b). Infection of the greater wax moth, *Galleria mellonella* (Linn.) with the nuclear polyhedrosis of the silkworm. *J. Insect Pathol.* **4**, 122–127.

Aizawa, K. (1963). The nature of infect'ons caused by nuclear-polyhedrosis viruses. *In* "Insect Pathology" (E. A. Steinhaus, ed.), Vol. 1, pp. 382–403. Academic Press, New York.

Aruga, H., and Nagashima, E. (1962). Generation-to-generation transmission of the cytoplasmic polyhedrosis virus of *Bombyx mori* (Linn.). *J. Insect Pathol.* **4,** 313–320.

Aruga, H., Hukuhara, T., Yoshitake, N., and Ayudaya, I. N. (1961). Interference and latent infection in the cytoplasmic polyhedrosis of the silkworm, *Bombyx mori* (Linn.). *J. Insect Pathol.* **3,** 81–92.

Bailey, L. (1962). Sacbrood virus of the larval honey bee (*Apis mellifera* Linn.). *Virology* **23,** 425–429.

Bailey, L. (1965). *in litt.*

Bailey, L., Gibbs, A. J., and Woods, R. D. (1963). Two viruses from the adult honey bee (*Apis mellifera* Linn.). *Virology* **21,** 390–395.

Bergold, G. H. (1958). Viruses of insects. *In* "Handbuch der Virusforschung" (C. F. Hallauer and K. F. Meyer, eds.), *Vol. 4,* pp. 60–142. Springer, Vienna.

Bird, F. T. (1959). Polyhedrosis and granulosis viruses causing single and double infections in the spruce budworm, *Choristoneura fumiferana* Clemens. *J. Insect Pathol.* **1,** 406–430.

Bird, F. T., and Whalen, M. M. (1953). A virus disease of the European sawfly *Neodiprion sertifer* (Geoffr.). *Can. Entomologist* **85,** 433–437.

Bolle, J. (1908). *Z. landwirtsch. Versuchsw. Oesterr.* **11,** 279–280.

Burnside, C. E. (1945). The cause of paralysis of honey bees. *Am. Bee J.* **85,** 354–355.

Conte, A. (1907). *Compt. Rend. Session Assoc. Franc. Advance. Sci.* **36,** 622–623.

Day, M. F. (1965). *Sericesthis* iridescent-virus infection of adult *Galleria. J. Invert Pathol.* **7,** 102–105.

Day, M. F., Farrant, J. L., and Potter, C. (1958). The structure and development of a polyhedral virus affecting the moth larvae. *Pterolocera amplicornis. J. Ultrastruct. Res.* **2,** 227–238.

Elmore, J. C., and Howland, A. F. (1964). Natural versus artificial dissemination of nuclear polyhedrosis virus by contaminated adult cabbage loopers. *J. Insect Pathol.* **6,** 430–438.

Escherich, K., and Miyajima, M. (1911). Studien über die Wipfelkrankheit der Nonne. *Naturw. Z. Forst- Landwirtsch.* **9,** 381–402.

Franz, J., Krieg, A., and Langenbuch, R. (1955). Untersuchungen über den Einfluss der Passage durch den Darm von Raubinsekten und Vögeln auf die Infektiosität insektenpathogener Viren. *Z. Pflanzenkrankh. Pflanzenschutz* **62,** 721–726.

Gershenson, S. M. (1959). Personal communication.

Gershenson, S. M. (1960). A study of a mutant strain of nuclear polyhedral virus of the oak silkworm. *Probl. Virol.* (*USSR*) (*English Transl.*) **6,** 720–725.

Harpaz, I., and Ben Shaked, Y. (1964). Generation-to-generation transmission of the nuclear polyhedrosis virus of *Prodenia litura* Fabr. *J. Insect Pathol.* **6,** 127–130.

Huger, A. (1963). Granuloses of insects. *In* "Insect Pathology" (E. A. Steinhaus, ed.) Vol. 1, pp. 531–575. Academic Press, New York.

Hukuhara, T. (1962). Generation-to-generation transmission of the cytoplasmic polyhedrosis virus of the silkworm, *Bombyx mori* Linn. *J. Insect Pathol.* 4, 132–135.

Jaques, R. P. (1962). The transmission of nuclear polyhedrosis virus in laboratory
populations of *Trichoplusia ni* (Hübner). *J. Insect Pathol.* 4, 433–445.

Krieg, A. (1955). Untersuchungen über die Polyhedrose von *Neodiprion sertifer*
(Geoffr). *Arch. Ges. Virusforsch.* 6, 163–174.

Maramorosch, K. (1951). A simple needle for micro-injections. *Nature* 167, 734.

Martignoni, M. E. (1955). Microinjector needle for determination of *per os* LD_{50}
of insect viruses. *Science* 122, 764.

Martignoni, M. E. (1957). Contributo alla conoscenza di una granulosi di *Eucosma
griseana* (Hübner) (Tortricidae: Lepidoptera) quale fattori limitante il pullu-
lamento dellinsetto nella Engadina alta. *Mitt. Schweiz. Zentral anst. Forstl.
Versuchsw.* 32, 371–418.

Martignoni, M. E. (1959). Preparation of glass needles for micro-injection. *J. Insect
Pathol.* 1, 294–296.

Martignoni, M. E., and Milstead, J. E. (1962). Trans-ovum transmission of the nuclear polyhedrosis virus of *Colias eurytheme* Boisd. through contamination of the
female genitalia. *J. Insect Pathol.* 4, 113.

Masera, E. (1954). Sul contenuto microbico intestinale del baco da seta e sull'e-
tiologia della flaccidezza. *Agr. Venezie* 8, 712–735.

Neilson, M. M. (1964). A cytoplasmic polyhedrosis virus pathogenic for a number of
lepidopterous hosts. *J. Insect Pathol.* 6, 41–52.

Neilson, M. M., and Elgee, D. E. (1950). The effect of storage on the virulence
of a polyhedrosis virus. *J. Insect Pathol.* 2, 165–171.

Ponsen, M. B., and deJong, D. J. (1964). A nuclear polyhedrosis of *Orthosia
incerta* (Hüfnagel) (Lepidoptera, Noctuidae). *J. Insect Pathol.* 6, 376–378.

Rivers, C. F. (1962). Personal communication.

Sidor, Ć. (1960). A polyhedral disease of *Chrysopa perla* L. *Virology* 10, 551.

Smirnoff, W. A. (1962a). A nuclear polyhedrosis of *Erannis tiliaria* (Harris) (Lepidoptera : Geometridae). *J. Insect Pathol.* 4, 393–400.

Smirnoff, W. A. (1962b). Trans-ovum-transmission of virus of *Neodiprion swainei*
Middleton (Hymenoptera : Tenthridinidae). *J. Insect Pathol.* 4, 192.

Smirnoff, W. A. (1963). Adaptation of a nuclear polyhedrosis virus of *Trichiocampus viminalis* (Fallén) to larvae of *Trichiocampus irregularis* (Dyar). *J.
Insect Pathol.* 5, 104–110.

Smirnoff, W. A. (1964). A nucleopolyhedrosis of *Operophtera bruceata* (Hulst.)
(Lepidoptera : Geometridae). *J. Insect Pathol.* 6, 384–386.

Smith, K. M. (1959). The use of viruses in the biological control of insect pests.
Outlook Agr. 2, 178–186.

Smith, K. M. (1960). Some factors in the use of pathogens in biological control
with special reference to viruses. *Rept. 7th Commonwealth Entomol. Conf.,
London, 1960* pp. 111–118.

Smith, K. M. (1963). The arthropod viruses. *In* "Viruses, Nucleic Acids and Cancer," pp. 72–84. Williams & Wilkins, Baltimore, Maryland.

Smith, K. M., and Rivers, C. F. (1956). Some viruses affecting insects of economic
importance. *Parasitology* 46, 235–242.

Smith, K. M., and Wyckoff, R. W. G. (1951). Electron microscopy of insect viruses. *Research (London)* **4**, 148–155.

Smith, K. M., Hills, G. J., and Rivers, C. F. (1959). Polyhedroses in neuropterous insects. *J. Insect Pathol.* **1**, 431–437.

Smith, K. M., Hills, G. J., and Rivers, C. F. (1961). Studies on the cross-inoculation of the *Tipula* iridescent virus. *Virology* **13**, 233–241.

Smith, K. M., Trontl, Z. M., and Frist, R. H. (1964). A note on a granulosis disease of a noctuid larvae. *Virology* **24**, 508–513.

Stairs, G. R. (1964). Infection of *Malacosoma disstria* Hübner with nuclear poly-hedrosis viruses from other species of *Malacosoma* (Lepidoptera, Lasiocampi-dae). *J. Insect Pathol.* **6**, 164–169.

Steinhaus, E. A. (1953). The susceptibility of two species of *Colias* to the same virus. *J. Econ. Entomol.* **45**, 897–898.

Steinhaus, E. A. (1954). Duration of infectivity of the virus of silkworm jaundice. *Science* **120**, 186–187.

Steinhaus, E. A. (1960). The duration of viability and infectivity of certain insect pathogens. *J. Insect Pathol.* **2**, 225–229.

Tanada, Y., and Chang, G. Y. (1962). Cross-transmission studies with three cyto-plasmic polyhedrosis viruses. *J. Insect Pathol.* **4**, 361.

Tanada, Y., Tanabe, A. M., and Reiner, C. E. (1964). Survey of the presence of a cytoplasmic polyhedrosis virus in field populations of the alfalfa caterpillar, *Colias eurytheme* Boisd., in California. *J. Insect Pathol.* **6**, 439–447.

Thompson, C. G., and Steinhaus, E. A. (1950). Further tests using a polyhedrosis virus to control the alfalfa caterpillar. *Hilgardia* **19**, 411–445.

Vago, C., and Croissant, O. (1959). Recherche sur la pathogénèse des viroses d'insectes. La libération des virus dans le tube digestif de l'insecte á partir des corps d'inclusion ingérés. *Ann. Épiphyties* **10**, 5–81.

White, G. F. (1917). Sacbrood. *U.S. Dept. Agr., Tech. Bull.* **431**, 1–54.

Wilson, F. (1960). The effectiveness of a granulosis virus applied to field popula-tions of *Pieris rapae* (Lepidoptera). *Australian J. Agr. Res.* **2**, 485–497.

CHAPTER VIII

Latent Viral Infections

A. Definitions

A good deal of confusion has arisen concerning the exact terms to be used to describe that phenomenon whereby an organism is infected with a virus but yet shows no apparent signs of infection. Different terms have been used to describe the same phenomenon or the same terms have been applied to two different conditions.

In order to clarify the situation a symposium was held in Wisconsin on "Latency and Masking in Viral and Rickettsial Infections" (Symposium, 1958), and a symposium on similar lines was held in Stockholm on the occasion of the 7th International Congress of Microbiology (Proceedings, 1958).

The definitions agreed on at the Wisconsin meeting are as follows:

Inapparent infection covers, at the host–parasite level, the whole field of infections which give no overt sign of their presence.

Subclinical can be used as an alternative, particularly in human medicine.

Latent infections are inapparent infections which are chronic, and in which a certain virus–host equilibrium is established. The adjective "latent" is best reserved to qualify "infection," the term "latent virus" being avoided.

Occult virus is used to describe the cases where virus particles cannot be detected and in which the actual state of the virus cannot as yet be ascertained. It is preferred to "masked," since this word has been used in a number of contradictory meanings.

Whenever it has been shown that viruses of animals or higher plants go through cycles as described for bacteriophages, the terms *provirus,*

146

vegetative virus and *infective virus* are appropriate for the corresponding stages. Infective virus is the fully formed virus particle.

A *moderate* virus is one growing in a cell while still permitting the latter's continued survival and multiplication; a *cytocidal* virus kills the cell. *Submoderate* covers intermediate cases; some viruses may be moderate in one cell-system and cytocidal in another. Lwoff (1958) points out that an infection, whether apparent or not, should not be recognized as viral until infectious particles have been detected and identified as a virus.

Latent virus infections are extremely frequent in most types of organisms, and this is particularly true of the viruses affecting plants and insects. Indeed in some insects latent infection is so common that entire populations of a species may be infected. Anyone who has tried to rear large numbers of caterpillars in captivity will be aware of the rapidity with which a virus disease can appear and spread through the insect stocks.

Latent virus infections may differ in the same insect species in different countries. For example, in Cambridge, England, it is extremely difficult to rear the larvae of *Pieris rapae*, "cabbage worms" in captivity. A latent granulosis develops and kills 90–100% of the larvae. In Pittsburgh, Pennsylvania, however, when the same species is reared in captivity, the larvae die in large numbers, not from a granulosis but from a cytoplasmic polyhedrosis.

Both these viruses are latent in the insects in a natural environment, but become virulent under conditions of captivity.

B. State of the Virus in Latent Insect Infections

Latent virus infection is of course intimately connected with the transovarial transmission of virus which we discussed in the previous chapter. We do not know what is the condition of latent viruses in insects, or what is the state of an occult virus. It may consist merely of naked nucleic acid, but there is no evidence available on this point at present.

Ring zones and small polyhedra have been observed in healthy stocks of *B. mori* larvae which had shown no indication of virus infection for several years (Roegner-Aust, 1949).

On several occasions both nuclear and cytoplasmic polyhedra have been observed in the imagoes of Lepidoptera. This, however, is probably an overt, rather than a latent, infection, since most of such infected adults usually die prematurely and it is not known whether any of them are capable of laying viable eggs. As regards the noninclusion-body viruses, particles of TIV have been observed in the head, behind the eyes, in adults of *Pieris brassicae,* the large white butterfly (Smith, 1963) (Fig. 42). Similarly the *Sericesthis* iridescent virus (SIV) has been found in the epithelial cells of adult *Galleria mellonella,* but in spite of this, transovum infection of SIV does not occur in the greater wax moth *G. mellonella* (Day, 1965).

On the occasions when viruses, either of the inclusion-body or noninclusion-body types, have actually been seen in the adult insect ovarial transmission of the virus does not always occur. It does happen, however, according to Hukuhara (1962) in the cytoplasmic polyhedrosis of the silkworm in which polyhedra were observed in the midgut epithelium of the adult moths.

In true latent infections no adverse effects of the virus upon the insect have been observed and a virus–host equilibrium seems to have been established.

C. Examples of Latent Infections

Many cases of latent virus infection in insects are known but it would serve no particular purpose to list them all. One or two examples only of each type of virus infection are therefore given.

Nuclear polyhedrosis viruses are frequently latent in *B. mori,* especially in countries which have long bred silkworms. In the spruce budworm, *Choristoneura fumiferana,* whole populations may carry a latent infection with a nuclear polyhedrosis virus. Similar latent infections have been observed in the larvae of the clothes moth *Tineola bisselliella* Hummel and in caterpillars of the puss moth *Cerura vinula* (Linn.). Latent infections with cytoplasmic polyhedrosis viruses occur in *B. mori,* tent caterpillars *Malacosoma* sp., in the winter moth *Operophtera brumata* Linn., in the pine looper *Bupalus piniarius* (Linn.) and, as previously mentioned, in *Pieris rapae.*

Adults of the alfalfa butterfly, *Colias eurytheme* Boisduval, from 12 localities covering almost the whole length of California, were found to be carrying a cytoplasmic polyhedrosis virus. The virus apparently occurred in these populations either in an occult or chronic condition (Tanada *et al.,* 1964).

Chronic latent infections with granuloses are common in *Pieris brassicae* Linn. and *P. rapae* Linn.

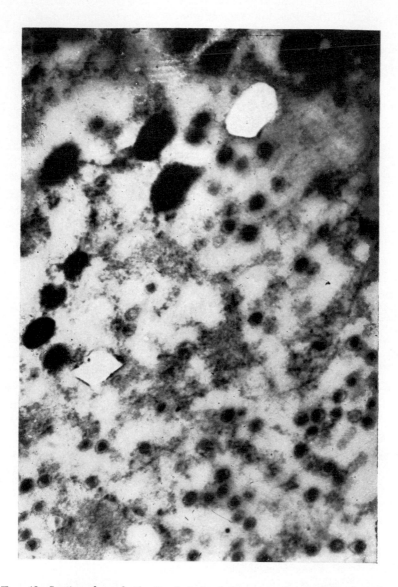

Fig. 42. Section through the head, behind the eyes of an adult white butterfly *Pieris brassicae* L. Note many particles of the *Tipula* iridescent virus; the large black bodies are pigment spots. Magnification, × 65,000.

Of the noninclusion-body viruses we have already mentioned TIV and SIV in adult *P. brassicae* and *G. mellonella,* respectively. The viruses of acute bee paralysis (ABPV) and chronic bee paralysis (CBPV) occur commonly in apparently healthy honeybees and also in bumblebees (Bailey *et al.,* 1963).

D. Conditions Governing Latency

Very little is known about the conditions which decide whether a virus infection will be latent or virulent. In plants where latent virus infections are frequent it has been suggested that latency may occur when the initial virus dose is very small (Bennett, 1958). Perhaps the same is true with insects for there is some evidence that the condition of latency is set up when the initial dose is small and is received in a late stage of larval development. Preliminary investigations have shown that weak concentrations of a nuclear polyhedrosis virus are nonfatal for older instars of sawfly larvae, *Neodiprion swainei* Middleton, since about 60% of them are able to develop into normal imagoes. These retain the faculty of transovum transmission. The sex-ratio of adults, the period of egg hatching, the period of incubation, and egg mortality remain practically unchanged (Smirnoff, 1962).

E. Induction of Latent Virus Infections

It is possible to stimulate latent virus infections in insects into virulence in a number of ways; these are called "stressors" by Steinhaus (1958a,b) who has made a preliminary study of stressors as a factor in insect disease.

Of natural stressors, crowding and temperature conditions are the commonest; the greater the number of insects confined in a given space, the greater the incidence of disease among them.

High temperatures do not appear to have much effect on latent virus infections, but it has been reported by some Japanese workers that exposure to high temperatures can induce insect virus infections, particularly in the silkworm. Exposure of silkworms to temperatures of 45° to 50°C for 10 minutes induced the development of a nuclear polyhedrosis (Kitajima, 1926). High-temperature treatment was effective in the induction of a cytoplasmic polyhedrosis in the silkworm (Hukuhara and Aruga, 1959), but low temperatures are more effective. When heat treatment is applied to portions of the larval body only, it seems that special organs or tissues situated in the anterior part of the larva play a part in the induction of polyhedroses. From these results, Hukuhara and Aruga (1959) consider that heat and cold treatments disturb the physiological conditions of larvae, culminating in an abnormal condition which in turn causes the transformation of the viruses from a noninfective to an infective state.

Steinhaus (1960) exposed a number of species of lepidopterous larvae to temperatures of 47°C for 10 minutes and to 51°C for 15 minutes. In none of these was there an increase in the incidence of polyhedroses over that occurring in the controls.

It has been shown by Japanese workers (Kurisu, 1955; Aruga, 1957; Aruga and Watanabe, 1961; Tanaka and Aruga, 1963) that low temperatures can stimulate into action a latent cytoplasmic polyhedrosis in the silkworm. Aruga and Hukuhara (1960) have experimented with various substances to induce nuclear and cytoplasmic polyhedroses in the silkworm, B. mori. They tested 18 chemicals by feeding them to fifth-instar larvae. Among these, 6 were found to be effective in the induction of cytoplasmic polyhedroses. These chemicals were sodium cyanide, sodium fluoride, arsenic acid, monoiodoacetic acid, sodium azide, ethylenediamine tetraacetic acid (EDTA), and its disodium salt. The last two chemicals induced cytoplasmic polyhedroses with high frequency when they were fed to fifth-instar larvae, but only with low frequency when fed to fourth-instar larvae.

The effects of several stressors have been studied on silkworms to which the cytoplasmic polyhedrosis virus was administered per os before or after the stress. Among stressors tested, those which increased the incidence of the cytoplasmic polyhedrosis were low temperatures (5°C for 3–5 hours), heat (37°C for 5 hours), formalin (0.01–1%), EDTA (powder), and acetic acid (0.1–0.5 M). On the other hand, calcium hydroxide (powder) decreased the incidence of the disease. There are two possible explanations for these effects: the stressors make easier the penetration of the virus into the cell, or the stressors accelerate the multiplication of the virus in the infected cell (Aruga et al., 1963).

Exposure to ether may be a factor in the induction of polyhedroses; in Peridroma larvae there appeared to be an increase in deaths in those larvae which had been exposed to etherization (Tanada, 1959; Steinhaus and Dineen, 1960).

Jaques (1960) exposed the cabbage looper, Trichoplusia ni, to vibration. He reported that the incidence of polyhedroses in the shaken larvae was 4 times that in the unshaken controls when both lots of larvae were fed low doses of polyhedra.

Aruga and Yoshitake (1961) found no significant increase in spontaneous polyhedroses when silkworms were exposed to ultraviolet radiation or X-rays. The percentage of nuclear and cytoplasmic polyhedroses was, however, markedly higher in larvae exposed to a double treatment of X-rays and low temperature, than in those treated with low temperature alone.

Density of insect population, or crowding, is a form of stressor which may activate latent virus infections, and with crowding goes another factor, the lack of sufficient food. This is shown in field populations of *Porthetria dispar;* a nuclear polyhedrosis frequently appears during mass infestations of this insect. The cytoplasmic polyhedrosis of the alfalfa caterpillar, *Colias eurytheme,* was discovered in insects subjected to the stress of crowding (Steinhaus and Dineen, 1960).

In some cases unsuitable food will provoke a latent virus infection but this does not always happen. Various workers have reported the onset of polyhedrosis in larvae of *B. mori* by changing the food plant from mulberry to *Scorzonera hispanica* Linn. (Ripper, 1915) or to *Maclura aurantiaca* Nutt. (Vago, 1951, 1955). On the other hand, Tanada (1956) was unable to induce the appearance of virus disease in an armyworm (*Pseudaletia unipunctata*) by changes in the food plant.

Another method of induction which, in the writer's experience, is frequently successful is inoculation of the larvae, by feeding, with a foreign virus. This has been demonstrated with larvae of the winter moth *Operophtera brumata* Linn., and of the pine looper *Bupalus piniarius* Linn. In each case, however, inoculation with a nuclear polyhedrosis virus provoked the development of a cytoplasmic polyhedrosis, and this seems to be what usually happens; the writer has not observed the reverse phenomenon, i.e., stimulation of a nuclear polyhedrosis by inoculation with cytoplasmic polyhedra. However, Krieg (1957) states that application of cytoplasmic polyhedra from *Dasychira pudibunda* Linn. (Lepidoptera) apparently stimulated into virulence a nuclear polyhedrosis in *Neodiprion sertifer* Geoffr., the European sawfly (Hymenoptera).

Suspensions of microsporidian spores have been shown capable of inducing active cytoplasmic polyhedrosis in the alfalfa caterpillar, even though the incidence of microsporidia may be low (Tanada, 1962). A combination of stressors is sometimes better than one acting alone. By the use of two stressors, a suspension of a microsporidian spore and of a granulosis virus, on the alfalfa caterpillar (*Colias eurytheme* Boisd.) certain individuals among adults collected from 12 localities in California were found to be carrying a cytoplasmic polyhedrosis virus, either in a latent or occult condition (Tanada *et al.*, 1964).

REFERENCES

Aruga, H. (1957). Mechanism of virus resistance in the silkworm, *Bombyx mori* II. On the relation between the nuclear polyhedrosis and the cytoplasmic polyhedrosis. *Nippon Sanshigaku Zasshi* **26**, 279–283. (English summary.)

Aruga, H., and Hukuhara, T. (1960). Induction of nuclear and cytoplasmic polyhedroses by feeding of some chemicals to the silkworm, *Bombyx mori* L. *Nippon Sanshigaku Zasshi* **29**, 44–49. (English summary.)

Aruga, H., and Watanabe, H. (1961). Difference in induction rate of polyhedroses by some treatments of inbred lines and their hybrids in the silkworm, *Bombyx mori* L. *J. Sericult. Sci. Japan* **30**, 36–42.

Aruga, H., and Yoshitake, N. (1961). Studies on the induction of nuclear and cytoplasmic polyhedroses by treating with X-rays and ultraviolet light in the silkworm, *B. mori* L. *Japan. J. Appl. Entomol. Zool.* **5**, 46–49 (in Japanese with English summary).

Aruga, H., Yoshitake, N., and Owada, M. (1963). Factors affecting the infection *per os* in the cytoplasmic polyhedrosis of the silkworm, *Bombyx mori* L. *Nippon Sanshigaku Zasshi* **32**, 41–50. (English summary.)

Bailey, L., Gibbs, A. J., and Woods, R. D. (1963). Two viruses from the adult honey bee (*Apis mellifera* Linn.). *Virology* **21**, 390–395.

Bennett, C. W. (1958). Masked plant viruses. *Proc. 7th Intern. Congr. Microbiol., Stockholm, 1958* pp. 218–223. Almqvist & Wiksell, Uppsala.

Day, M. F. (1965). *Sericesthis* iridescent virus infection of the adult *Galleria*. *J. Invert. Pathol.* **7**, 102–105.

Hukuhara, T. (1962). Generation-to-generation transmission of the cytoplasmic polyhedrosis virus of the silkworm, *Bombyx mori* L. *J. Insect Pathol.* **4**, 132–135.

Hukuhara, T., and Aruga, H. (1959). Induction of polyhedrosis by temperature treatment in the silkworm (*Bombyx mori* L.). *Nippon Sanshigaku Zasshi* **28**, 235–241. (English summary.)

Jaques, R. P. (1960). The ecology of the polyhedrosis of the cabbage looper, *Trichoplusia ni* (Hübner). Ph.D. Thesis, Cornell University, Ithaca, New York.

Kitajima, E. (1926). High temperature and grasserie. *Sangyo-Shinpo* **393**, 148–154.

Krieg, A. (1957). "Toleranzphänomen" und Latenzproblem. *Arch. Ges. Virusforsch.* **7**, 212–219.

Kurisu, I. (1955). Studies on virus diseases in the silkworm. I. On the mid-gut polyhedrosis. *Bull. Kumamota Sericult. Expt. Sta.* **6**, 48–56.

Lwoff, A. (1958). "Recent progress in microbiology. Symp. IV. Latent and masked virus infections." *Proc. 7th Intern. Congr. Microbial., Stockholm, 1958* p. 211. Almqvist & Wiksell, Uppsala.

Proceedings (1958). "Recent progress in microbiology. Symp. IV. Latent and masked virus infections." *Proc. 7th Intern. Congr. Microbiol., Stockholm, 1958* pp. 453. Almqvist & Wiksell, Uppsala.

Ripper, M. (1915). Bericht über die Tätigkeit der K. K. Landwirtschaftlichchemischen Versuchsstation in Görz im jahre 1914. *Z. Landwirtsch. Versuchsw. Deut.-Oesterr.* **18**, 12–15.

Roegner-Aust, S. (1949). Der Infektionsweg bei der Polyederepidemic der Nonne. *Z. Angew. Entomol.* **31**, 3–37.

Smirnoff, W. A. (1962). Transovum-transmission of virus of *Neodiprion swainei* Middleton (Hymenoptera: Tenthredinidae). *J. Insect Pathol.* **4**, 192.

Smith, K. M. (1963). The arthropod viruses. *In* "Viruses, Nucleic Acids, and Cancer," p. 76. Williams and Wilkins, Baltimore.

Steinhaus, E. A. (1958a). Stress as a factor in insect disease. *Proc. 10th Intern. Congr. Entomol., Montreal, 1956* Vol. 4, pp. 725–730. Intern. Congr. Entomol., Ottawa, Canada.

Steinhaus, E. A. (1958b). Crowding as a possible stress factor in insect disease. *Ecology* **39**, 503–514.

Steinhaus, E. A. (1960). Notes on polyhedroses in *Peridroma, Prodenia, Colias, Heliothis,* and other lepidoptera. *J. Insect Pathol.* **2**, 327–333.

Steinhaus, E. A., and Dineen, J. P. (1960). Observations on the role of stress in a granulosis of the variegated cutworm. *J. Insect Pathol.* **2**, 55–65.

Symposium (1958). *Symposium on Latency and Masking in Viral and Rickettsial Infections, Wisconsin, 1957* pp. 202. Burgess, Minneapolis, Minnesota.

Tanada, Y. (1956). Some factors affecting the susceptibility of the armyworm to virus infections. *J. Econ. Entomol.* **49**, 52–57.

Tanada, Y. (1959). Microbial control of insect pests. *Ann. Rev. Entomol.* **4**, 277–302.

Tanada, Y. (1962). Effect of a microsporidian spore suspension on the incidence of cytoplasmic polyhedrosis in the alfalfa caterpillar, *Colias eurytheme* Boisd. *J. Insect Pathol.* **4**, 495–497.

Tanada, Y., Tanabe, A. M., and Reiner, C. E. (1964). Survey of the presence of a cytoplasmic polyhedrosis virus in field populations of the alfalfa caterpillar, *Colias eurytheme* Boisd, in California. *J. Insect Pathol.* **6**, 439–447.

Tanaka, S., and Aruga, H. (1963). The effect of cold treatment on the polyhedrosis virus infections in the silkworm, *B. mori* L. *Nippon Sanshigaku Zasshi* **32**, 226–231. (English summary.)

Vago, C. (1951). Phénomène de "latentia" dans une maladie à ultravirus des insectes. *Rev. Can. Biol.* **10**, 299–308.

Vago, C. (1955). Facteurs alimentaires et activation des viroses latentes chez les insectes. *Proc. 6th Intern. Congr. Microbiol., Rome, 1953* pp. 556–564.

CHAPTER IX

Tissue Culture of Insect Viruses

Introduction

One of the fundamental characteristics of viruses is their inability to grow in a cell-free or synthetic medium; they can only multiply in a living susceptible cell. However, since the cell, and not the whole organism, is the nurse to a virus it follows that viruses can be grown in cells which are themselves being cultivated in a nutritive medium.

This is what is meant by the tissue culture of viruses, and it consists of two types. One type uses explants, pieces of tissue—in insects, whole organs—and the other employs only dispersed tissues or loose cells. It is cell culture with which we are mainly concerned here.

Development of tissue-culture techniques has been one of the most important advances in recent years insofar as the viruses attacking man and the higher animals are concerned. From these advances has risen the hope of controlling such diseases as measles, poliomyelitis, and the common cold.

Although tissue-culture methods have been known for over 40 years, it was not until 1925 that a virus, vaccinia, was shown to have increased by about 50,000 times when cultured in rabbit testicular tissue.

A number of discoveries have helped to give the tissue culture of viruses its present importance; the main impetus was supplied by the urgent necessity of finding a vaccine for poliomyelitis.

One important discovery was the recognition that viruses growing in cells produce changes which are easily recognized and are characteristic of the virus in question. This allows for a rapid diagnosis of some virus infections, especially polio virus. The recognizable cytopathic effects (CPE) in cell cultures of polio virus have thus done away with the

155

necessity of using large numbers of monkeys for tests or even as a source of virus.

All the procedures in the technique of tissue culture must, of course, be free of bacterial contamination. The discovery and development of antibiotics have proved of inestimable value to tissue-culture methods, since by their addition to the medium bacterial contamination can be controlled without effect on the virus.

Another great step forward was the development of very uniform cultures, consisting of a single sheet or monolayer of cells. This was made possible by the treatment of cell suspensions with trypsin. Recently Dulbecco and his colleagues (Dulbecco, 1952) have developed this method still further by the "plaque technique." They found that the spread of virus through the culture could be prevented by overlaying the cell monolayer with nutrient agar immediately after inoculation of the virus into the cells. Under these conditions only direct spread of the virus is possible, and this results in the development of plaques which become visible to the naked eye. Since each plaque represents one infective unit of virus, an accurate method of virus assay is obtained.

As might be expected, a cell culture consists of a heterogeneous collection of cells which do not react in a uniform manner to a given virus. Better results are therefore obtained when the cell culture is a "clone" in which all the cells are derived from a single cell. One of the best known of these cell lines is a strain known as HeLa cells. These came originally from a human cancer, cervical carcinoma tissue, and have been adopted for use in many laboratories.

Unfortunately the tissue culture of insect and plant viruses has lagged far behind the tissue culture of viruses affecting the higher animals. This was inevitable, so far as insect viruses are concerned, since the whole study of insect virology is so recent.

The first attempts at insect tissue culture were made by Goldschmidt (1915) who carried out some experiments on spermatogenesis *in vitro*. Since that time various workers have experimented with insect-tissue culture but it is only quite recently that some tangible progress has been made.

In their survey of the culture of insect tissues, Day and Grace (1959) point out two main requirements for success in this field. First, the insect should be one in which the relevant tissues grow during larval life by increase in the number of cells. Secondly, a comprehensive analysis of the hemolymph of the chosen insect should be available. Much time and effort have been wasted owing to the use of unsuitable media, and not much progress has been made. However, S. S. Wyatt (1956) and G. R. Wyatt *et al.* (1956) made a chemical study of silkworm hemolymph and introduced a medium which is a great improvement, and this with various modifications, has been used successfully with several species of Lepidoptera.

As regards the best type of insect tissue to use, Grace (1954) found that the blood cells of silkworms formed a syncytium within 20 hours and this survived for 6 days.

However, ovarian tissue seems to be the most successful; Trager (1935) was the first to demonstrate that the inner envelope of the ovarian follicle of caterpillars provided the cells which grew best in culture. Subsequent work with the silkworm has confirmed that this tissue grows better in culture than any other tested.

Glucose is the principal source of energy in most of the media used for the vertebrate culture media, and S. S. Wyatt (1956) found that a mixture of glucose, fructose, and sucrose in the proportions of 4 : 2 : 1 gave the best results.

Certain substances which are present in insect hemolymph are essential ingredients in culture media. These include high concentrations of amino acids and a high proportion of organic acids; S. S. Wyatt (1956) used a mixture of malate, ketoglutarate, succinate, and fumarate which resulted in a marked stimulation of growth.

Insects require vitamins, especially some of the B vitamins necessary for vertebrates.

The temperature generally utilized for insect tissue culture has been room temperature or between 24° and 26°C. (Day and Grace, 1959). [See also the review by Maramorosch (1962).]

Since ovarian tissue has proved the most successful method of insect cell culture, some details of the procedure are given as described by Martignoni (1962). In the larva and pupa of Bombycidae and Saturniidae the ovaries are found dorsally in the region of the fifth and sixth abdominal segments. They are excised and transferred to a balanced salt solution or culture medium and then cut into pieces of about 1 mm³. The structure of the immature ovariole differs from that of the mature ovariole by the presence of an intermediate layer of disconnected oval cells (Jones and Cunningham, 1961; Trager, 1935). A syncytial outer epithelial sheath encloses the ovariole. Separated by a membrane, an intermediate layer of cells is found beneath the outer sheath in the immature ovariole. Beneath this intermediate layer and separated from it by a thin membrane, follows the follicular epithelium which encloses the oöcyte and the trophocytes. It is the cells of this intermediate layer which become predominant from the explant.

We have mentioned earlier that trypsin is widely used in the breakup or disaggregation of vertebrate tissue. There is some evidence that trypsin may be rather destructive of insect cells, although it has been used successfully by Grace (1962). It may be that there is a variable response to trypsin by cells of different origin.

For the disaggregation of tissues from the larvae of *Peridroma saucia* (Hübner), Martignoni *et al.* (1958) found that extract from the hepato-pancreas and crop of the snail *Helix aspersa* Müller was better than trypsin. Vago and Bergoin (1963) have made cultures of fibroblasts from ovarial tubes of *Lymantria dispar* in liquid medium as well as in plasma clot and in monolayers. These cultures allowed the *in vitro* study of the pathogenesis of the nuclear and cytoplasmic polyhedroses of this insect. In addition, infection of the *Lymantria* fibroblasts with viruses from other insects was also obtained (Fig. 43).

Before we proceed to a discussion of the various techniques and media used in the cell culture of insects, there is another aspect of the subject which must be briefly mentioned. In Chapter XI an account is given of the relationships of insects with plant viruses, and it is shown how in certain cases the borderline between plant and insect viruses becomes indistinct and doubtful. In other words, some plant viruses transmitted by leafhoppers (Cicadellidae) actually multiply in the tissues of the leafhopper vector, thus opening up a new and fascinating field of study.

Although the multiplication of several viruses in the leafhopper has been demonstrated (Black, 1959), the actual site of virus multiplication has not been determined.

The culture of tissues and various organs of cicadellid leafhoppers, the vectors of many plant viruses, is now being studied by Maramorosch and his co-workers (Hirumi and Maramorosch, 1963, 1964; Mitsuhashi and Maramorosch, 1964; Hirumi, 1963).

As regards the tissue culture of the actual organs of leafhopper vectors, success has been claimed by Hirumi and Maramorosch (1963). Testes, ovaries, gut, brain, Malpighian tubules, and salivary glands of two species of cicadellid leafhoppers were successfully cultured in a variety of media. Best results were obtained in media containing minerals, incorporated in Vago's medium number 22, plus newborn calf

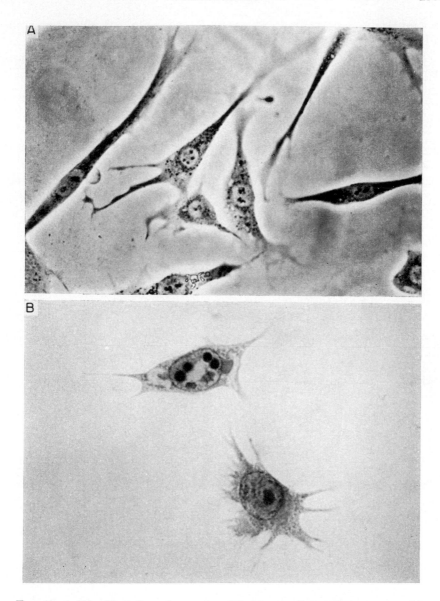

FIG. 43. A, Fibroblasts from the ovaries of *Lymantria dispar*. Phase contrast. Magnification, × 650. B, Fibroblasts from *Lymantria dispar* infected with the nuclear polyhedrosis from *Bombyx mori* the silkworm. The developing nuclear polyhedra can be seen. Magnification, × 1100. (C. Vago and M. Bergoin, 1963.)

serum, leafhopper serum and extracts, and serum of the armyworm (*Leucania unipuncta*).

By this means it may be possible to ascertain which of these organs is the site of virus multiplication.

A. Techniques and Media

After the pioneer experiments of Trager (1935), one of the most successful culture media was that proposed by S. S. Wyatt (1956), who used silkworm blood plasma and a high concentration of amino acids particularly rich in lysine, histidine, and serine in order to balance the medium with the amino acid composition of the silkworm blood plasma. The composition of Wyatt's medium is shown in Table IV.

TABLE IV

Tissue-Culture Medium[a] for *Bombyx mori*[b]

Component	Concentration, mg/100 ml	Component	Concentration, mg/100 ml
NaH$_2$PO$_4$	110	DL-Alanine	45
MgCl·6H$_2$O	304	β-Alanine	20
MgSO$_4$·7H$_2$O	370	L-Proline	35
KCl	298	L-Tyrosine	5
CaCl$_2$	81	DL-Threonine	35
		DL-Methionine	10
Glucose	70	L-Phenylalanine	15
Fructose	40	DL-Valine	20
Sucrose	40	DL-Isoleucine	10
		DL-Leucine	15
L-Arginine HCl	70	L-Tryptophan	10
L-Histidine	250	L-Cystine	2.5
L-Aspartic acid	35	Cysteine HCl	8
L-Asparagine	35		
L-Glutamic acid	60	Malic acid	67
L-Glutamine	60	α-Ketoglutaric acid	37
Glycine	65	Succinic acid	6
DL-Serine	110	Fumaric acid	5.5
		Antibiotics	(as required)

[a] pH, 6.5.

[b] From S. S. Wyatt, 1956.

Martignoni and Scallion (1961) have used hemocyte monolayers obtained from larvae of *Peridroma saucia* (Hübner). The hemocytes were obtained from 2–4-day-old sixth-instar larvae (last larval stage). The larvae were anesthetized with ethyl ether for about 10 minutes, surface-sterilized with 0.2–0.4% aqueous solution of Hyamine 10-X for 5 minutes with agitation and then allowed to dry between folded sterile paper towels. The larvae are drained of the germicide for about 2 minutes. One proleg of each larva is amputated and the blood is allowed to drain into 3–4 ml of culture medium in a 12-ml conical graduated tube. About 15–20 larvae can be handled as a group.

The composition of the final culture medium is 15% insect blood, 15% fetal bovine serum, and 70% maintenance medium. The composition of the latter is shown in Table V.

After gentle mixing by pipette the cell suspension is distributed in Petri dishes, each containing one 22×22-mm No. 1 cover glass. The volume used is 3.5 ml for 60-mm glass Petri dishes, or 2 ml for the 35-mm polystyrene dishes. The cover glasses can be used for permanent preparations after the culture is discarded.

The Petri dishes must be placed without delay in the incubator in the following gas mixture, 96% N_2, 3% CO_2 and 1% O_2. The medium is replaced every third day. Good results can be obtained without insect blood plasma by using a replacement medium consisting of 30% fetal bovine serum and 70% of the medium 26C (Table V).

The following method of preparing the maintenance medium (26C) is given by Martignoni and Scallion (1961). The medium is prepared with stock solutions; $100 \times$ concentrates of the L-amino acids and of the vitamins are commercially available. A 100-ml volume of the medium is prepared as follows:

(1) Place in a beaker 90 ml of ascorbic acid-free, buffered (pH 6.4) balanced salt solution (BSS).

(2) Add 1 ml each of $100 \times$ concentrates of the amino acids and vitamins; usually L-glutamine is stored as a separate $100 \times$ concentrate.

(3) Add 1 ml of $100 \times$ concentrate of penicillin and streptomycin.

(4) Add and allow to dissolve 70 mg glucose, 40 mg fructose, and 40 mg trehalose.

(5) Add and allow to dissolve 100 ml L-ascorbic acid and carefully neutralize, under constant stirring, using molar NaOH solution, to pH 6.45; about 1 ml will be required.

(6) Bring volume up to 100 ml with BSS.

TABLE V

MEDIUM 26C[a,b]

Component	Concentration, mg/100 ml	Component	Concentration, mg/100 ml
KCl	626.2	L-Tryptophan	0.4
CaCl$_2$	255.3	L-Tyrosine	1.8
MgCl$_2$·6H$_2$O	366.0	L-Valine	2.4
H$_2$(HPO$_3$)	114.9	L-Glutamine	30.0
NaOH	106.8		
		L-Ascorbic acid	100.0
Glucose	70.0	Biotin	0.1
Fructose	40.0	Choline	0.1
d(+)-Trehalose	40.0		
		Folic acid	0.1
L-Arginine	2.1	Nicotinamide	0.1
L-Cystine	1.2	Pantothenic acid	0.1
L-Histidine	0.8	Pyridoxal	0.1
L-Isoleucine	2.6	Thiamine	0.1
L-Leucine	2.6	Riboflavin	0.01
L-Lysine	2.6	Isoinositol·2H$_2$O	0.18
L-Methionine	0.8		
L-Phenylalanine	1.6		
L-Threonine	2.4	Antibiotics	(as required)

Gas phase, % by volume	
Oxygen	1
Carbon dioxide	3
Nitrogen	96
Freezing-point depression, °C	0.60

[a] After Martignoni and Scallion (1961); amino acids and vitamins according to Eagle (1955, 1959).

[b] pH, 6.45.

(7) Sterilize by filtration (Millipore filter, type HA); store at 4°C. Do not use after 5 days' storage.

The composition of the balanced salt solution is shown in Table VI.

A considerable contribution to the subject has been made by Grace (1962) who established 4 strains of cells from ovarian tissue of *Antheraea eucalypti,* and the following account is from his paper. The cultures from which the cells were developed consisted of cells from ovaries of diapausing pupae which had been dissociated in 0.25% trypsin (Difco trypsin, 1 : 250) in Puck's Saline A. After dissociation, the cells from 4

TABLE VI
Balanced Salt Solution[a]

Component		Quantity, gm
No.	Name	
1	NaCl	6.8
2	KCl	0.4
3	CaCl₂	0.2
4	MgSO₄	0.1
5	NaH₂PO₄	0.125
6	NaHCO₃	2.2
7	Dextrose	1.0
8	Water[b]	—

[a] After Earle (1934).

[b] Procedure: In preparation in lots of 1 liter, items 1, 6, and 7 are weighed out and dissolved in about 500–700 ml water. To this solution are added items 2 and 5 in the form of stock solutions. The mixture is then made up to 1000 ml and approximately saturated with CO_2 by bubbling the gas through it for a few minutes. The solution can be sterilized by filtration, under 8–14 lb pressure, through a Mandler, Seitz, or Berkefeld filter.

ovaries were placed in a Porter's flask with 0.9 ml of insect tissue culture medium. All cultures were incubated at 27°–29°C.

The medium was changed every 7–14 days for the first 7 months, but after the cells began to multiply rapidly it became necessary to change the medium every 4–5 days, and subcultures were made at 8–10 day intervals. The cells were removed from the bottle for subculturing in the following way. The medium was removed from the bottles and the cells were washed once in Puck's Saline A. Then 2.0 ml of 0.1% trypsin in Puck's Saline A were added and the cells incubated at 25°C. As soon as a few cells had come free from the glass, 1.0 ml of insect tissue-culture medium was added and gently pipetted to free the cells from the glass. After centrifugation at 200 g for 2 minutes the supernatant was removed and the cells resuspended in fresh medium. Approximately 50,000–100,000 cells/ml were placed in a 2-oz flat medicine bottle with 30 ml of medium. The medium used was a modification of that described by S. S. Wyatt (Table IV), chlolesterol being added as well as 10 members of the vitamin B complex, and the blood concentration reduced from 10% to 5%. The composition of Grace's medium is shown in Table VII.

TABLE VII
INSECT TISSUE–CULTURE MEDIUM[a]

Component		Concentration,	Component		Concentration,
Group	Name	mg/100 ml	Group	Name	mg/100 ml
A. Salts			C. Sugars		
	$NaH_2PO_4 \cdot 2H_2O$	114		Sucrose (gm)	2.668
	$NaHCO_3$	35		Fructose	40.0
	KCl	224		Glucose	70.0
	$CaCl_2$ (separate)	100			
	$MgCl_2 \cdot 6H_2O$	228			
	$MgSO_4 \cdot 7H_2O$	278	D. Organic acids		
				Malic acid	67
B. Amino acids				α-Ketoglutaric acid	37
	L-Arginine HCl	70			
	L-Aspartic acid	35		Succinic acid	6
	L-Asparagine	35		Fumaric acid[c]	5.5
	L-Alanine	22.5			
	β-Alanine	20			
	L-Cystine HCl	2.5	E. Antibiotics		
	L-Glutamic acid	60		Penicillin "G"	
	L-Glutamine	60		(Na salt)	3
	L-Glycine	65		Streptomycin sulfate	10
	L-Histidine	250			
	L-Isoleucine	5	F. Vitamins		
	L-Leucine	7.5		Thiamine HCl	0.002
	L-Lysine HCl	62.5		Riboflavin	0.002
	L-Methionine	5		Calcium pantothenate	0.002
	L-Proline	35		Pyridoxine HCl	0.002
	L-Phenylalanine	15		p-Aminobenzoic acid	0.002
	DL-Serine	110		Folic acid	0.002
	L-Tyrosine[b]	5		Niacin	0.002
	L-Tryptophan	10		Isoinositol	0.002
	L-Threonine	17.5		Biotin	0.001
	L-Valine	10		Choline chloride	0.002

[a] After Grace (1962).
[b] Dissolved in 1 N HCl.
[c] Neutralize organic acids with KOH.

METHOD OF PREPARATION OF GRACE'S MEDIUM

A, C, and D were each dissolved in 20 ml water, B in 30 ml water. The calcium chloride was dissolved separately in a small amount of water and was added after A, B, C, and D were thoroughly mixed. The vitamins were then added, and the pH adjusted to 6.5 with potassium hydroxide. The streptomycin and penicillin were added just before the medium was sterilized by passage through a sintered glass filter (porosity 5). After sterilization, 19-ml aliquots were taken and made up to 20.0 ml with 1.0 ml of heat-treated (60°C for 5 minutes) insect plasma from diapausing pupae of *A. eucalypti.*

Grace (1962) points out that in establishing these cell lines certain principles have emerged. Whenever possible the medium should contain plasma from the insect, the tissues of which are being cultured. Osmotic pressure, pH, and ionic ratios of the blood are important, and these factors should be altered in the medium when necessary. If it is not practicable to obtain a supply of homologous plasma, it is suggested that silkworm plasma be substituted. Cells from various species of Lepidoptera have grown well in media containing silkworm plasma.

Vago and Bergoin (1963) have cultured successfully tissues of the larvae or pupae of *Lymantria dispar.* They found that the pH of the hemolymph of this species was similar to that of the silkworm, *B. mori* (6.4–6.6). Cultures were made from the female gonads of the larvae or the ovarial tubes of the pupae. Larvae of *L. dispar* are very suitable for the extraction of the gonads, since their light color causes the latter to stand out from the surrounding tissue. Cultures of fibroblasts from ovarial tubes of *L. dispar* were obtained in liquid medium as well as in plasma clot and in monolayers (Fig. 43). The composition of the medium used by Vago and Bergoin is given in Table VIII.

Various pupal tissues of the Monarch butterfly *Anosia plexippus* L. have been cultivated *in vitro* in order to study the differences in cell type and growth potential of different organs and tissues (Hirumi and Maramorosch, 1964b).

Pupae were surface-sterilized in 70% ethanol for 5 minutes and then quickly flamed over a gas microburner. The pupae were covered with balanced salt solution (Table VI) and various organs excised under a magnification of 45 ×. Excised brain, optic lobes, subesophageal ganglion, circulatory organ, ovarian tissues, Malpighian tubules, and pupal wing tissues were placed separately in sterile Petri dishes containing the culture medium. This consisted of a mixture of 20 ml of modified Vago's *Bombyx mori* medium (Hirumi, 1964), 20 ml of Morgan's synthetic medium TC 199, and 13 ml of fetal bovine serum. TC Yeastolate was added in an amount equal to 0.1% of the medium. One hundred units per milliliter of penicillin and of streptomycin was also added to the mixture.

TABLE VIII

INSECT TISSUE–CULTURE MEDIUM[a]

Ingredient	Quantity
Distilled water,[b] ml	99
Phenol red, 2%, ml	1
$NaH_2PO_4 \cdot H_2O$, gm	0.120
$MgCl_2 \cdot 6H_2O$, gm	0.300
$MgSO_4 \cdot 7H_2O$, gm	0.400
KCl, gm	0.300
$CaCl_2 \cdot 2H_2O$, gm	0.100
Trehalose, gm	0.150
Lactalbumin hydrolyzate, gm	1.100
KOH, 5%[c]	[c]
Penicillin, IU	60,000
Streptomycin, mg	2

[a] After Vago and Bergoin (1963).
[b] To the physiological solution is added 12% of calf serum and 3% serum of *Bombyx mori* or *Lymantria dispar*.
[c] Adjust to pH 6.4.

Excised organs were cut into small pieces with sterile, fine dissecting knives. Sitting-drop cultures were prepared in the V-H (Vago-Hirumi) tissue culture flasks (Fig. B). Five or six tissue fragments, in a drop of culture medium, were placed on the cover-glass surface of each flask by means of capillary-tip pipettes. Excess medium was removed with a small pipette, thus allowing the tissue fragments to come gently into contact with the cover glass. They were then allowed to become partially dry and later a few drops of fresh culture medium were placed gently on the semidry tissue fragments. To keep the inside of the V-H flasks moist, a few drops of culture medium were used to cover the interior of the top of each flask with a thin layer of medium (Hirumi and Maramorosch, 1964b).

To find the proper material for cultivation *in vitro* of cells of leafhopper vectors of plant viruses, embryonic tissues of the six-spotted aster leafhopper (*Macrosteles fascifrons*) were tested during early developmental stages. Growing cells were only obtained from the stage of blastokinetic movement. This stage was the only one that yielded growing cells and thus is the only one suitable for cultivation *in vitro*.

This stage which is obtained on the seventh and eighth day after

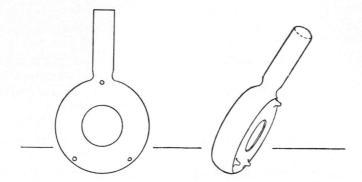

Fig. B. Vago-Hirumi tissue culture flask, actual size. Dorsal view at left shows position of hole and legs: side view at right indicates degree of neck constriction. (From Hirumi, 1963.)

oviposition can be identified by observing the development of the embryo. It occurs during the movement of the pigmented eye disk from the posterior to the anterior position in the embryo.

The culture medium used was the same as that employed for the culture of the tissues of monarch butterfly larvae, except that only 6 ml of fetal bovine serum was used instead of 13 ml.

The eggs were collected in batches of 8 or 10 and immersed in the culture medium. They were then removed, washed in balanced salt solution, and placed for 60 seconds in 70% ethanol. Next, they were washed again in the balanced salt solution and immersed for surface sterilization for 10 minutes in 0.1% Hyamine 238 g.

After further washing in salt solution, the eggs are passed 3 times through fresh culture medium on Maximow slides. The embryo is dissected out with a sterile knife under a dissecting microscope, washed 3 times in culture medium, and treated with a 0.02% trypsin solution. "Sitting-drop" cultures were prepared by placing 0.2 ml of medium with suspended trypsinized tissue fragments on the cover glass of each V-H tissue culture flask (Fig. B).

Embryonic tissues during early developmental stages, on the first, third, and fifth day after oviposition, and in late developmental stages (that is, tenth and eleventh day) as well as ovarian tissues of fifth-instar nymphs and of adults yield no growing cells (Hirumi and Maramorosch, 1964a).

B. Results Achieved

Trager (1935), who was the first to grow the virus of the silkworm nuclear polyhedrosis in tissue culture, observed the development of the polyhedra in cultures containing cells from the lining of the ovariole of the silkworm. Grace (1958), studying latent virus infections in insects, found that a latent virus infection in tussock moth larval tissue could be stimulated into action by a drastic change in the culture medium. He also observed the development, in his tissue cultures, of a previously latent cytoplasmic polyhedrosis when the cells were inoculated with a nuclear polyhedrosis virus. This is a similar phenomenon to that shown by Smith and Rivers (1956) in the larva itself. Caterpillars of the winter moth, *Operophthera brumata*, and the pine looper, *Bupalus piniarius*, when inoculated with a foreign nuclear polyhedrosis virus, developed a cytoplasmic polyhedrosis hitherto latent in the larvae.

Using the techniques already described, Martignoni and Scallion (1961) infected larval hemocytes with a nuclear polyhedrosis virus. Infection *in vitro* produced the typical sequence of changes which have been observed *in vivo*. The cytopathic effect is particularly evident in the large flattened nuclei of the amebocytes with a loss of basophilia, central condensation of the chromatin mass, and formation of the polyhedra within the nucleus. If the cells are fixed and stained during the stage of polyhedral formation, when residues of the central chromatin mass are still visible, the inclusions are strongly eosinophile, and infected cells stand out from their healthy neighbors by the bright red color of the inclusion bodies.

Very successful results in the cultivation of monolayers of insect cells have been claimed by Chinese workers (Gaw Zan-Yin *et al.*, 1959). They used Trager's solution A and 10% of healthy silkworm serum at an initial pH of 6.7. The silkworm serum was obtained by bleeding full-grown silkworms aseptically from the leg, centrifuging the blood at 2000 rpm for 10 minutes and adding penicillin and streptomycin to the final solution. The cells were separated by trypsinization using Bacto trypsin.

The monolayers obtained by the Chinese workers were obtained from various insect tissues, male and female gonads, tracheae, muscle, intestine and silk glands. The cells obtained were epithelial cells or fibroblasts.

The cultures were infected with a nuclear polyhedrosis virus from the silkworm and studies of the cytological changes were made. These changes appear to have been very similar to what have already been observed in the insect itself. The nuclei became greatly enlarged and changed to horseshoe, ring, or half-moon shapes as polyhedra began to develop. Finally, the nuclei burst; this was followed by disruption of the cell itself, liberating the polyhedra into the culture medium.

Gaw Zan-Yin and his co-workers claim that the success of this tissue culture renders unnecessary the raising of the silkworm itself for virus studies.

Vago and Bergoin (1963) have studied the pathogenesis of nuclear and cytoplasmic polyhedral viruses of *Lymantria dispar* in fibroblast cultures from ovarial tubes of this insect. In addition the development of viruses from other insects was studied. The viruses of nuclear polyhedroses from *Bombyx mori* and *Antheraea pernyi*, and occasionally a granulosis virus from *Pieris brassicae*, would also infect the fibroblasts from *L. dispar*. A fibroblast infected with the nuclear polyhedrosis virus of *Bombyx mori* is shown in Fig. 43.

Perhaps the most sophisticated approach to this problem so far has been made by Bellett and Mercer (1964) and Bellett (1965b), who have studied the multiplication of *Sericesthis* iridescent virus (SIV) in cell cultures from *Antheraea eucalypti* Scott.

We have already described the qualitative experiments of Bellett and Mercer on the multiplication of SIV in Chapter VI so they will not be dealt with further.

Bellett (1965a) has developed an *in vitro* assay by counts of cells stained by fluorescent antibody.

The methods of cell culture were similar to those used by Bellett and Mercer (1964) except that Grace's (1962) medium (see Table VII) was supplemented with bovine plasma albumin (BPA, 10 mg/ml) and the hemolymph concentration was reduced to 2%. Infected cultures were maintained in Grace's medium with two-thirds of the normal BPA and hemolymph content to prevent growth of the cells. The fluorescent antibody was prepared as described by Bellett and Mercer (1964).

The following method which is similar to that of Easterbrook (1961) was used for counting the cells stained by fluorescent antibody. A washed suspension of 10^5 cells was dried on a slide, fixed, and stained as described previously, and the proportion of stained cells was determined in 4 unselected fields; a total of 200–400 cells was usually counted. When less than 1% of the cells were stained, the total number of fluorescent cells in the sample was counted directly.

REFERENCES

Bellett, A. J. D. (1965a). The multiplication of *Sericesthis* iridescent virus in cell cultures from *Antheraea eucalypti* Scott. II. An *in vitro* assay for the virus. *Virology* 26, 127–131.

Bellett, A. J. D. (1965b). The multiplication of *Sericesthis* iridescent virus in cell cultures from *Antheraea eucalypti* Scott. III. Quantitative experiments. *Virology* 26, 132–141.

Bellett, A. J. D., and Mercer, E. H. (1964). The multiplication of *Sericesthis* iridescent virus in cell cultures from *Antheraea eucalypti* Scott. I. Qualitative experiments. *Virology* 24, 645–653.

Black, L. M. (1959). Biological cycles of plant viruses in insect vectors. *In* "The Viruses" (F. M. Burnet and W. M. Stanley, eds.), Vol. 2, pp. 157–185. Academic Press, New York.

Day, M. F., and Grace, T. D. C. (1959). Culture of insect tissues. *Ann. Rev. Entomol.* 4, 17–38.

Dulbecco, R. (1952). Production of plaques in monolayer tissue cultures by single particles of an animal virus. *Proc. Natl. Acad. Sci. U.S.* 38, 747–752.

Eagle, H. (1955). The minimum vitamin requirements of the L and HeLa cells in tissue culture, the production of specific vitamin deficiencies, and their cure. *J. Exptl. Med.* 102, 595–600.

Eagle, H. (1959). Amino acid metabolism in mammalian cell cultures. *Science* 130, 432–437.

Earle, W. R. (1934). A technique for the adjustment of oxygen and carbon dioxide tensions, and hydrogen ion concentration, in tissue cultures planted in Carrel flasks. *Arch. Exptl. Zellforsch. Gewebezucht,* 16, 116–128.

Easterbrook, K. B. (1961). The multiplication of vaccinia virus in suspended KB cells. *Virology* 15, 404–416.

Gaw Zan-Yin, Liu Nien Tsui, and Zia Tien un (1959). Tissue culture methods for cultivation of virus grasserie. *Acta Virol.* (*Prague*) 3, Suppl., 55–60.

Goldschmidt, R. (1915). Some experiments on spermatogenesis *in vitro*. *Proc. Natl. Acad. Sci. U.S.* 1, 220–222.

Grace, T. D. C. (1954). Culture of insect tissues. *Nature* 174, 187–188.

Grace, T. D. C. (1958). Effect of various substances on growth of silkworm tissues *in vitro*. *Australian J. Biol. Sci.* 2, 407.

Grace, T. D. C. (1962). Establishment of four strains of cells from insect tissues grown *in vitro*. *Nature* 195, 788–789.

Hirumi, H. (1963). An improved device for cultivating cells *in vitro* and for observations under high-power phase magnification. *Contrib. Boyce Thompson Inst.* 22, 113–116.

Hirumi, H., and Maramorosch, K. (1963). Cultivation of leaf hopper (Cicadellidae) tissues and organs *in vitro*. *Ann. Epiphyties* 14, 77–79.

Hirumi, H. and Maramorosch, K. (1964a). The *in vitro* cultivation of embryonic leaf hopper tissues. *Exptl. Cell Res.* 36, 625–631.

Hirumi, H., and Maramorosch, K. (1964b). Tissue culture of the Monarch butterfly, *Anosia plexippus* L. *Contrib. Boyce Thompson Inst.* 22, 259–268.

Jones, B. M., and Cunningham, I. (1961). Growth by cell division in insect tissue culture. *Exptl. Cell Res.* **23**, 386–401.

Maramorosch, K. (1962). Present status of insect tissue culture. *Proc. 11th Intern. Congr. Entomol., Vienna, 1960* Vol. 2, pp. 801–807.

Martignoni, M. E. (1962). Insect tissue culture: A tool for the physiologist. *Proc. 23rd Biol. Colloq., Oregon State Univ., 1962* pp. 89–110.

Martignoni, M. E., and Scallion, R. J. (1961). Preparation and uses of insect hemocyte monolayers *in vitro*. *Biol. Bull.* **121**, 507–520.

Martignoni, M. E., Zitcer, E. M., and Wagner, R. P. (1958). Preparation of cell suspensions from insect tissue for *in vitro* cultivation. *Science* **128**, 360.

Mitsuhashi, J., and Maramorosch, K. (1964). Leafhopper tissue culture: embryonic, nymphal, and imaginal tissues from aseptic insects. *Contrib. Boyce Thompson Inst.* **22**, 435–460.

Smith, K. M., and Rivers, C. F. (1956). Some viruses affecting insects of economic importance. *Parasitology* **46**, 235–242.

Trager, W. (1935). Cultivation of the virus of grasserie in silkworm tissue culture. *J. Exptl. Med.* **61**, 501–513.

Vago, C., and Bergoin, M. (1963). Développement des virus a corps d'inclusion du lépidoptère *Lymantria dispar* en cultures cellulaires. *Entomophaga* **8**, 253–261.

Wyatt, G. R., Loughheed, T. C., and Wyatt, S. S. (1956). The chemistry of insect hemolymph. Organic components of the hemolymph of the silkworm, *Bombyx mori* and two other species. *J. Gen. Physiol.* **39**, 853–868.

Wyatt, S. S. (1956). Culture *in vitro* of tissue from the silkworm *Bombyx mori* L. *J. Gen. Physiol.* **39**, 841–852.

CHAPTER X

Further Aspects of the Relationships between Insects and Viruses

A. Mixed Infections, Interference and Synergism

Just as composite virus diseases occur in plants and the higher animals, so infection of insects with two unrelated viruses is common enough. It would serve no useful purpose to list all the cases of double virus infections which have been described, but a few examples may be cited. Double infection of *B. mori* with both nuclear and cytoplasmic polyhedroses have been recorded by Smith and Xeros (1953a,b) and by Smith *et al.* (1953). Similar complex infections seem to be common also in Japan (Aizawa, 1963). Double infections with a nuclear polyhedrosis and a granulosis have also been frequently recorded. The following, among others, have been reported by Steinhaus (1957): *Euxoa segetum* (Schiffermüller), *Pieris rapae* (Linn.), *Pseudaletia unipuncta, Sabulodes caberata* (Guenée), *Estigmeme acrea* Drury, and *Autographa californica* Speyer. Double infections of a dipterous larva, *Tipula paludosa* (Linn.) with its nuclear polyledrosis and the noninclusion-body virus, known as the *Tipula* iridescent virus (TIV) have been found on several occasions by the writer.

The phenomena of mixed virus infections and interference have been much studied in plants (Kassanis, 1963). This interference may be such that in the case of two closely related viruses, the virus which enters the plant first, sometimes, but not with every virus, prevents the entrance, or at least the multiplication, of the second virus, provided the first virus has had time to become systemic in the plant. This is known as cross protection. In Chapter XI where the relationships of plant viruses with insects are discussed, it is shown that one strain of aster yellows inhibits the replication in the *insect vector* of a second related strain of the same virus. In other words, where plant viruses multiply in the insect vector the same type of cross protection may exist in the insect as in the plant.

The question then arises: Is there a similar type of interference when an insect becomes infected with two similar insect viruses? To achieve cross protection between like viruses in plants, it is necessary for one virus to be inoculated first and to be followed later by the second or challenge virus. A somewhat similar, though less clear-cut result, has been described by Aruga *et al.* (1963a). Silkworms are susceptible to cytoplasmic polyhedrosis virus from the pine caterpillar, *Dendrolimus spectabilis* (Butler), if inoculated in the first instar. The pine-caterpillar virus, when fed to larvae 1 to 5 days before or simultaneously with the silkworm virus, apparently interfered with infection by the silkworm virus. On the other hand, no interference was apparent when the pine-caterpillar virus was fed to larvae 1 or 2 days after the feeding of the silkworm virus. Moreover, the simultaneous feeding of the silkworm and the pine-caterpillar viruses did not greatly affect the incidence of infection by the latter virus. Apparently the simultaneous feeding of the silkworm and the pine-caterpillar viruses to first-instar silkworms resulted in relatively few cases of mixed infection. The two viruses could be differentiated in the bodies of affected silkworms by the shape of the polyhedra, tetragonal in outline for the silkworm virus and hexagonal for the pine-caterpillar virus. The mechanism underlying this kind of interference is not clear, unless it is that the pine-caterpillar virus develops more quickly or is slightly more virulent to the silkworm than its own cytoplasmic virus.

Experimental results more in line with the cross protection in plant viruses and in insect vectors have been reported by Aruga *et al.* (1963b). In this case two cytoplasmic polyhedrosis viruses of the silkworm were used, one giving rise to icosahedral polyhedra and the other to hexahedral polyhedra. It was found that when the dose of the challenge virus remained constant, the higher the dose of the first-administered virus, the lower the percentage of the occurrence of cytoplasmic polyhedrosis by the challenge virus. This looks like a case of first-come, first-served, and it may be that a longer time interval before the application of the challenge virus might serve the same purpose as a larger dose of the first-administered virus. In other words the first virus would have more time to occupy the available multiplication sites.

The interactions of two cytoplasmic polyhedrosis viruses in the silkworm and the alfalfa caterpillar, *Colias eurytheme* (Boisd.) have been studied (Tanada and Chang, 1964). In these two larvae the interactions of the viruses varied with the insect species, the dosage of virus inoculum, the sequence and interval in which each virus was fed to the same larva, and the age of the host larva. The interaction in the silkworm larva appeared to be a mutual coexistence of the two viruses. In the case of the alfalfa caterpillar, its virus interfered with infection by the silkworm virus. The extent of the interference varied with the concentrations of the two viruses and with the sequence in time of feeding of one virus before the other. When, in the mixture of the two viruses the concentration of the polyhedra of the *Colias* virus was equal to or greater than that of the *Bombyx* virus, the former virus interfered with infection by the latter. Interference also occurred in the *Colias* larva when, at the same polyhedra concentration, the *Colias* virus was fed prior to or shortly after the *Bombyx* virus.

In other words, in the silkworm its own virus seems to increase infection by the *Colias* virus. In the *Colias* larva its own virus interferes with infection by the silkworm virus. Interference between two strains of a nuclear polyhedrosis virus in *Hyphantria cunea* has been observed (Aruga *et al.*, 1961).

So far we have only discussed the interaction of viruses of the same group, even if they were from different host species. But larvae are sometimes infected at the same time with two different types of viruses, a cytoplasmic polyhedrosis and a granulosis or a nuclear polyhedrosis and a granulosis. The reactions in these two combinations seem to depend upon three factors, (1) an advantage in time, i.e., if one virus infects first; (2) a more rapid development, and (3) competition for the same sites of replication. If larvae of *Pieris brassicae* (Linn.) are infected simultaneously with a cytoplasmic polyhedrosis virus and a granulosis virus both types of inclusion bodies are found in the tissues but the larvae die of the granulosis. Here is a case where one virus, that of the cytoplasmic polyhedrosis, develops in the cells of the gut while the other, that of the granulosis, develops in the fat body. Since there is no competition and since the granulosis virus develops more rapidly, this virus causes the death of the insect (Vago, 1959).

The results, however, are different when the two infecting viruses are those of a granulosis and a nuclear polyhedrosis. In this case there is competition for the same tissue, the fat body, and a more rapid rate of development for the nuclear polyhedrosis virus, 4–13 days as compared to 10–34 days for the granulosis virus. In consequence the larvae of the armyworm *Pseudaletia unipuncta* (Haworth), when thus infected, die of the nuclear polyhedrosis and not the granulosis (Tanada, 1959).

Spruce budworm larvae, *Choristoneura fumiferana* Clemens, are susceptible to a nuclear polyhedrosis virus which multiplies in the nuclei of tracheal matrix, fat, hypodermal, and blood cells, and to a less virulent granulosis virus which multiplies in the cytoplasm of the same cells. Prior infection by one virus interferes with infection by the second. In general, it appears that a cell will accept either virus but not both. Adjacent cells are frequently infected with different viruses, but usually blocks of cells are infected with the same virus. To produce double infection, an advantage must be given to the granulosis virus either in time or size of dose (Bird, 1959).

A somewhat similar result has been observed in double infections in the silkworm larvae with the nuclear polyhedrosis viruses of *B. mori* and *Samia cynthia pryeri*. Each type of polyhedra was formed in a cell nucleus, but both kinds of polyhedra were not observed in the same nucleus. However, large triangular polyhedra, which were different from either of the two types of polyhedra inoculated, were formed in a few nuclei of cells (Ishikawa and Asayama, 1961). There does not seem to be any information concerning these triangular polyhedra as to whether they might possibly be produced by a hybrid of the two viruses.

The occurrence of two viruses together in the same insect sometimes has a synergistic effect; in other words the two viruses acting in unison produce a more severe disease than each acting alone. Tanada (1959) reports that a granulosis virus has a synergistic effect when accompanying a nuclear polyhedrosis virus in the armyworm *Pseudaletia unipuncta* (Haworth). Larvae are most susceptible when fed both viruses at the same time. An initial feeding of the granulosis virus followed by one of the nuclear polyhedrosis virus increases the susceptibility of the armyworm more than the reverse procedure. The heat-inactivated granulosis virus is still capable of enhancing the virulence of unheated nuclear polyhedrosis virus, but heat-inactivated nuclear polyhedrosis virus has no apparent effect on the unheated granulosis virus, suggesting that the synergistic effect rests with the granulosis virus.

B. Immunity and Resistance

It had long been thought that the adult insect was immune to infection with the polyhedrosis viruses, but evidence is now accumulating that this is not so. It is frequently possible to produce infected adults by infecting lepidopterous larvae with virus in a late developmental stage. Such adults may die prematurely or suffer reduced fecundity. Smith (1963) observed cytoplasmic polyhedra in the midgut of several adult species; most of these survived for only a short time.

Neilson (1965) has studied the effects of a cytoplasmic polyhedrosis on the adults of four species of Lepidoptera: *Alsophila pometaria* (Harris), *Nymphalis antiopa* (Linn.), *Operophtera brumata* (Linn.), and *Paleacrita vernata* (Peck). Since this was a cytoplasmic polyhedrosis virus polyhedra were observed, as in the larvae, only in the midgut cells of infected adults. The average size of diseased specimens was smaller than that of healthy individuals in all species, and tumor-like structures were found in a large proportion of both male and female diseased *A. pometaria*, *O. brumata*, and *P. vernata* adults.

In addition, the wings of most diseased males of *A. pometaria* and *O. brumata* were malformed and the virus interfered with oocyte development in the females.

In all four species, virus infection greatly reduced the reproductive ability of the adults.

In the foregoing the virus of a cytoplasmic polyhedrosis was transmitted indirectly to the adult moths via the caterpillars. It is possible, however, to infect adult insects, derived from healthy larvae, directly by inoculation with a nuclear polyhedrosis virus. A nuclear polyhedrosis was produced experimentally in adults of a noctuid moth *Peridroma saucia* (Hübner). The changes characteristic of this type of virus disease developed in the same tissues as would be infected in the larva, i.e., the fat body, tracheal matrix, and the epidermis. Diseased moths lived for a considerably shorter period than did the healthy moths (Martignoni, 1964).

There is some evidence of an acquired immunity in silkworms against their nuclear polyhedrosis. Yamafuji *et al.* (1958) first inoculated the silkworms with the virus, and then injected them at intervals with an antiserum against the virus. Aizawa (1962) found that the larva of the greater wax moth, *Galleria mellonella* (Linn.), was susceptible to a nuclear polyhedrosis of the silkworm. The virus is thought to be a *Galleria*-adapted strain of the silkworm virus and it was neutralized by an antiserum prepared against the nuclear polyhedrosis from the silkworm, but not by an antiserum prepared against the silkworm cytoplasmic polyhedrosis virus. Therapy by antiserum against silkworm nuclear polyhedrosis virus is thus effective. It is not clear to what this immunity is due since insects have not been shown to produce antibodies. An antibody is generally considered to be a modified blood globulin formed in response to an antigenic stimulus; it is capable of combining specifically with the corresponding antigen. As there is no proof that insects possess globulin similar to the gamma globulin of mammals, it is inadvisable to assume that they can produce antibodies (Stephens, 1963).

There are a few examples on record of resistance in insects to their virus diseases; it seems necessary to assume that resistance must exist, otherwise there would be no survivors in the epizootics which frequently occur when populations of insects become excessive. Glaser (1915), quoted by Martignoni and Schmid (1961), concluded

If racial or acquired immunity does not exist among certain individuals of the gypsy moth, it is difficult to understand how any of these insects escape death under certain conditions in the field.

Bird and Elgee (1957) suggested that the European sawfly developed some resistance to a nuclear polyhedrosis. This was based on recent increases in sawfly numbers from a very low to a relatively high population level without a subsequent rapid increase in larval mortality from disease.

Ossowski (1958, 1960) found that equal concentrations of a nuclear polyhedrosis virus from different localities varied in their virulence towards the wattle bagworm, *Kotochalia junodi* (Heyl.). From this, Ossowski concluded that the virus existed in different strains of varying virulence. It could, however, be said with equal truth that the bagworms showed a difference in susceptibility to infection. This is supported by some work on the resistance of certain stocks of *Pieris brassicae* larvae to infection with a granulosis virus. That stocks of *P. brassicae* do differ in their susceptibility to infection has been shown by David (1957), Rivers (1959), and Sidor (1959). The larvae of a culture of *P. brassicae*, previously shown to be resistant to granulosis (David and Gardiner, 1960), are still resistant, in comparison with another stock, after 4 years and 36 generations. This is true also when comparison is made with virus derived from a third stock. A different race of the insect, *P. brassicae* race *cheiranthi*, was imported from the Canary Islands; this insect which feeds on nasturtium plants, *Tropaeolum majus*, Linn., in preference to cabbage, brought with it a granulosis virus of its own.

Of three stocks of *P. brassicae* larvae studied, the resistant stock retained its resistance against its own granulosis virus and virus from another British stock. This latter stock was itself more resistant than the stock from the Canary Islands which was equally susceptible to virus from the British stock or its own virus.

By a direct comparison, at equal concentrations of virus derived from the *cheiranthi* stock and from the second British stock, it was shown that the granuloses from these two sources were equally virulent and consequently that one from a distant source was not more effective against a local larva than its own virus (David and Gardiner, 1965a,b,c).

A study of the degree of susceptibility to two virus diseases was carried out in California in 1958–1959, with four populations of two species of Lepidoptera, *Phryganidia californica* Packard (nuclear polyhedrosis), and *Pieris rapae* (Linn.) (granulosis). The two populations of *P. californica*, a native species with a long association with its pathogen, showed differences in susceptibility and degree of heterogeneity, while the two populations of *P. rapae*, a species recently introduced into California, gave identical responses. A certain degree of resistance was detected in the population of *P. californica* which was known to have a large amount of stability (a so-called umbral population) while an unstable population of the same insect in the phase of increasing density, was found to be less resistant and more heterogeneous (Martignoni and Schmid, 1961).

Certain conditions may govern susceptibility or resistance to virus infection. There seems little doubt that the early instars of larvae are more susceptible than older larvae. For example, silkworms can be infected with a cytoplasmic polyhedrosis from the pine caterpillar, *Dendrolimus spectabilis* (Butler), but only if inoculated in the first instar (Aruga *et al.*, 1963a).

Environmental conditions, especially temperature, play an important part in the infection process. So far as the inclusion-body diseases are concerned, Tanada (1953) reported that some larvae of *Pieris rapae* (Linn.) survived infection from granulosis when reared at 36°C. The larvae of the sawfly *Diprion hercyniae* (Hartig) when reared continuously at 29.4°C (Bird, 1955) and those of the cabbage looper, *Trichoplusia ni* (Hübner) and the corn earworm, *Heliothis zea* (Boddie) (Thompson, 1959) resisted infection by their respective nuclear polyhedrosis viruses. Apparently there is no question here of the virus being inactivated inside the insect, as happens to the aster-yellows virus inside its insect vector (see chapter XI). Because of this Thompson concludes the action of the high temperature is on the mode or mechanism of infection.

The resistance of *Galleria mellonella* to the *Tipula* iridescent virus (TIV) at high temperatures has been studied.

At 23°–25°C, the larvae of *G. mellonella*, which had been inoculated in the hemocoele with TIV died of infection. At a temperature above 30°C, the virus-inoculated larvae survived to become adults. It is considered that as the virus was inoculated directly into the hemocoele, the resistance of the larvae at high temperatures was due to the destruction of the virus, or the prevention of virus multiplication by immune and non-immune host reactions, and not to the failure of the virus to penetrate into the hemocoele (Tanada and Tanabe, 1965).

C. Virus Strains and Mutations

One of the characteristics that viruses share with microorganisms is the ability to mutate and this, of course, is a commonplace with plant viruses and viruses of the higher animals. Virus strains and mutations in the plant viruses can be recognized in various ways, slight differences in symptoms, slight chemical changes, or alterations in their serology. In the insect viruses techniques are less well-developed and recognition of virus strains is not so straightforward. In the inclusion-body diseases a clue to the appearance of a new virus strain or mutation is found in the shape of the polyhedral crystals, since this is a function of the virus and not of the host cell (Gershenson, 1960).

By selection through 13 generations Aizawa *et al.* (1961) obtained a silkworm strain of a nuclear polyhedrosis virus which was resistant to induction by cold treatment (see Chapter VIII).

A variant of the silkworm nuclear polyhedrosis virus was thought to have been obtained by alternating passage through chick embryos and silkworm pupae. The polyhedra formed by the variant were tetragonal instead of the more usual hexagonal shape (Aizawa, 1961). In examining the shapes of the polyhedra in a nuclear polyhedrosis it is sometimes possible to observe some aberrant forms; if these are isolated and propagated, they remain constant in shape. These aberrant shapes are presumably caused by a mutant virus, although there seems to be little information on any differences in the virus itself.

A similar phenomenon has been observed in a granulosis disease of *Choristoneura fumiferana* (Clemens). In this instance the aberrant granules were cube-shaped instead of the normal polyhedral type. Propagation of these aberrant forms proved that they retained their cubic shape (Stairs, 1964).

It might perhaps be said that the *Sericesthis* iridescent virus was a strain of the *Tipula* iridescent virus since they have so many characters in common. The recent discovery of a similar virus affecting verte-brates (*in litt.;* Waterson, 1965) may afford opportunity for some inter-esting comparisons.

Very little effort seems to have been exerted to produce mutants of insect viruses by means of irradiation or other means. Prolonged so-journ in an unusual host may sometimes cause changes in a virus. With plant viruses this ocurs in a strain of tobacco mosaic virus (TMV) orig-inally isolated from beans. When this virus is cultivated in tobacco slight chemical changes take place and these are reversed on the return of the virus to the bean which was its original host; prolonged sojourn in either host is not necessary (Bawden, 1958).

As we have already mentioned in discussing the iridescent viruses, it would be interesting to propagate the *Sericesthis* iridescent virus in *Tip-ula paludosa,* the original host of TIV, to see if the differences between the two viruses are maintained.

D. Serology of Insect Viruses

The investigation of the serological relationships of the insect viruses has been much neglected and it is only recently that a start has been made. This is a fruitful field of study and should yield information which would be useful in many ways. The number of apparently distinct viruses would almost certainly be reduced and light would probably be thrown onto the question as to how far the large numbers of cytoplasmic polyhe-drosis viruses are one and the same. A sound system of nomenclature and classification also needs to be based on, among other things, the sero-logical relationships of the different viruses. It would be interesting also to ascertain if there existed any kind of serological relationship between some of the insect viruses and those viruses which are capable of multi-plying in both insects and plants (Chapter XI).

Much of the serological work done so far relates to the rod-shaped vi-ruses of the nuclear polyhedroses and granuloses; this is understandable enough since they are so much easier to purify than the cytoplasmic polyhedrosis viruses. One of the difficulties of obtaining antisera to the inclusion-body viruses is the necessity to make sure that the antiserum is to the virus particle and not to the outer membranes or inclusion-body protein.

Krywienczyk et al. (1958) have shown that viruses from the nuclear polyhedroses and the granuloses belong to two serologically distinct groups, since no cross reaction occurs between them, and this was to be expected since they are distinct in other ways. They also showed that there was a strong cross reaction between viruses from closely related hosts, but with polyhedrosis viruses from less closely related hosts there was only a weak cross reaction.

In a further paper (Krywienczyk and Bergold, 1960a) the serological relationships of 17 insect viruses from 4 continents were investigated by the complement-fixation technique. On the whole, the results support the grouping of the viruses into polyhedroses and granuloses, although one granulosis virus from *Recurvaria milleri* Busck gives detectable cross reaction with 6 polyhedrosis viruses. The polyhedrosis viruses from the Hymenoptera, two in number, give strong cross reactions only within their own group and are probably closely related if not identical.

In a paper published shortly after the above, Krywienczyk and Bergold (1960b) consider that when intact virus is injected into guinea pigs, antibodies are formed mainly against the membranes. They suggest that serological studies of intact insect viruses so far reported are in reality studies of the antigenic structure of the virus membranes, and not of the virus nucleoprotein. It would therefore be of interest to study the serological relationships of the DNA from the different nuclear polyhedrosis viruses and also between the DNA and RNA viruses.

The same workers (Krywienczyk and Bergold, 1960c) have studied the serological relationship between the inclusion-body proteins of Lepidoptera and Hymenoptera. They find that the proteins belong to three serologically distinct groups. Two of these groups, the nuclear-polyhedron proteins and the granulosis proteins from Lepidoptera, are the same as those established for the corresponding viruses. The granulosis protein from *R. milleri*, however, showed, like its virus, a higher cross reactivity with polyhedron proteins than with the other granulosis proteins. The third group is represented by the nuclear-polyhedron proteins of the sawflies (Hymenoptera).

Stobbart (cited in Smith et al., 1961) has carried out complement-fixation and gel-diffusion tests on the *Tipula* iridescent virus from 3 different hosts in the Lepidoptera, *Porthetria dispar* (Linn.), *Pieris brassicae* (Linn.), and *Nymphalis io* (Linn.). From these tests the TIV from all three hosts appears identical. Additional tests carried out with TIV from its natural host, the fly larva *Tipula paludosa* Meig., showed the virus to be the same.

Some studies on the serological properties of *Sericesthis* iridescent virus (SIV) have been made (Day and Mercer, 1964) using antisera produced in rabbits. An aqueous extract from a single larva showed a precipitation zone in agar only if it were not diluted to more than 10 ml. The antiserum had a titer of 10^{-2} in similar tests.

This antiserum did not react against intact SIV particles prepared either in buffer or in distilled water. If, however, the pH of the SIV solution was raised to 10.5, then antigens were detectable, showing identity with those of the infected larvae. No such antigen appeared after treatment of the virus at pH 3.0.

Aqueous extracts of infected larvae centrifuged at 39,000 rpm for 1 hour contained a soluble antigen, identical with an antigen obtained for the purified virus either by treatment at pH 10.5 or by sonic disintegration. This antigen is destroyed by alcohol and by acetone, but not by 75°C for 5 minutes.

A comparison between the serological responses of SIV and those of TIV showed marked differences between the two.

E. Artificial Feeding Media

It may be suggested that in the future, insect tissue-culture methods will be developed sufficiently to render unnecessary the propagation of large numbers of caterpillars for virus research. In research on the viruses of the higher animals, of course, this has already happened. Tissue culture has, for example, replaced the use of large numbers of monkeys for investigations on poliovirus and vaccine production.

Similarly the development of synthetic or semi-synthetic feeding media may eliminate the necessity for growing large numbers of plants for insect feeding. This would be a great advantage in cities or where it is not easy to propagate plants in winter time. Furthermore the medium can be sterilized, thus avoiding possible contamination of the insects by virus on the host plants or by other pathogens.

Synthetic feeding would also be useful in the production of large quantities of virus for such uses as chemical studies on the virus or for application in biological control. If the insect can be bred all the year round, this is obviously an advantage; a method for accomplishing this with the large white butterfly, *Pieris brassicae* (Linn.), has been developed (David and Gardiner, 1952). This is a useful insect for producing granulosis virus or *Tipula* iridescent virus (TIV) in quantity, as the caterpillar is quite large and gives a good yield of either virus.

Ignoffo (1963) has developed a successful method for mass-rearing the cabbage looper, *Trichoplusia ni* Hübner, on a semi-synthetic diet and the following account is taken from his paper.

The diet described here is basically the pink-bollworm medium devised by Vanderzant and Reiser (1956a,b). To this basic medium is added cotton-leaf stock, ascorbic acid, and, if mold contaminants increase, additional formaldehyde. Aureomycin is also included to eliminate yeast growth and further to reduce mold and bacterial contamination. The following formula will make about 3.6 liters of finished medium (see Table IX).

The cotton-leaf stock is prepared by drying cotton leaves in a dry-air sterilizer at 120°C for 1–2 hours and then grinding the leaves to a fine powder with a pestle and mortar.

TABLE IX

SEMI-SYNTHETIC DIET FOR *Trichoplusia ni* HÜBNER[a]

Ingredient	Quantity
	in milliliters
Distilled water	3,100
Methyl p-hydroxybenzoate	
(15% w/v in 95% ethyl alcohol)	36
Choline chloride (0.1 gm/ml water)	36
Potassium hydroxide 4 M	18
Formaldehyde (0.1 gm/ml water)	15
Vitamin stock	6
	in grams
Casein	126
Sucrose	126
Wheat germ	108
Cotton-leaf stock	54
Agar	90
Salts, Wesson's	36
Alphacel	18
Ascorbic acid	15
Aureomycin	0.5

[a] Ignoffo (1963).

The vitamin stock contains 600 mg niacin, 600 mg calcium pantothenate, 300 mg riboflavin, 150 mg each of thiamin, pyridoxin, and folic acid, 12 mg biotin, and 1.2 mg of vitamin B^{12} in 100 ml of water.

PREPARATION OF MEDIUM AND REARING CONTAINERS

The agar is placed in 2220 ml of water and dissolved in a boiling-water bath. The casein is blended with 880 ml of water and 18 ml of potassium hydroxide in a 1-gallon capacity commercial blender. All solids, with the exception of ascorbic acid and aureomycin, are then added and blended into a homogeneous mixture. The methyl p-hydroxybenzoate, choline chloride, formaldehyde, and vitamin solutions are successively added with continuous blending. The dissolved agar is added next and blended further. The ascorbic acid and aureomycin are added last, and the entire mixture is blended for about 2 minutes or until a homogeneous color is obtained. The hot medium is poured into rearing jars, capped with metal lids, tilted to distribute the medium over half the inner surface of the jar and then cooled in a horizontal position. After cooling, the jars are labeled and stored cap-side down.

Rearing Jars. The rearing container is a $5 \times 2\frac{1}{8}$-inch sterile glass jar filled with 35 ml of the medium. The medium is distributed over half the inner surface of the jar and provides about 8 sq. inches of feeding area. A 4×4-inch fine-mesh bolting cloth is placed over the open end of the jar after infestation and held in place by a No. 8 rubber band. A metal cap is screwed on the jar immediately after the medium is poured in and is kept on until the medium is infested with larvae. This technique has been used by Ignoffo (1964a) for the production of a nuclear polyhedrosis virus from *T. ni.*

The preceding medium has been modified by David and Gardiner (1965a) for the successful rearing of *Pieris brassicae* larvae. Details of their medium (Table X) and technique follow.

The dried cotton leaf in Ignoffo's medium (about 1.4% by weight) is replaced by dried cabbage-leaf powder. This is prepared by drying cabbage leaves in thin layers in a ventilated oven at 105°C for 15–20 minutes. The leaves can then be ground by hand with a roller and sieved through a 60-mesh sieve, but it is quicker to have them ground in a small mill fitted with a 0.5-mm-mesh screen. The veterinary-grade aureomycin soluble powder contained 25 gm chlortetracycline hydrocloride per pound.

TABLE X

MEDIUM FOR *Pieris brassicae* L.[a]

Ingredient		Quantity
Group	Component	
(a)	Distilled water, ml	110
	KOH, 4 M, ml	1.8
	Casein (light white soluble), gm	12.6
(b)	Sucrose, gm	12.6
	Wheat germ (Bemax), gm	10.8
	Cabbage (dried powder), gm	5.4
	Salt mixture,[b] gm	3.6
	Whatman Chromedia cellulose powder	
	(CF11 grade), gm	1.8
(c)	Choline chloride (10% solution), ml	3.6
	Methyl p-hydroxybenzoate	
	(15% in 95% Ethanol), ml	3.6
	Formaldehyde (10% W/V solution), ml	1.5
	Vitamin stock,[c] ml	0.8
(d)	Distilled water, ml	200
	Agar (fine Japanese powder), gm	9
(e)	L-Ascorbic acid, gm	1.5
	Aureomycin (veterinary grade), gm	0.5

[a] After David and Gardiner (1965a).

[b] The composition of the salt mixture (in gm) is: $CaCO_3$, 120; K_2HPO_4, 129; $CaHPO_4 \cdot 2H_2O$, 30; $MgSO_4 \cdot 7H_2O$, 40.8; NaCl, 67; $FeC_6H_5O_7 \cdot 6H_2O$, 11, KI, 0.32; $MnSO_4 \cdot 4H_2O$, 2.0; $ZnCl_2$, 0.10; $CuSO_4 \cdot 5H_2O$, 0.12 (Beckman *et al.*, 1953).

[c] The composition of the vitamin stock (in mg) is: nicotinic acid, 600; calcium pantothenate, 600; riboflavin (B_1), 300; aneurine hydrochloride (B_2), 150; pyridoxine hydrochloride (B_6), 150; folic acid, 150; D-biotin, 12; cyanocobalamine (B_{12}), 1.2; all in 100 ml water.

The ingredients listed in the Table under (a) are placed in a blender of 800-ml capacity, and thoroughly mixed together. The mixed solids (b) are then added with further blending. The solutions (c) are next added, separately, while the blender is running. Meanwhile the agar solution (d) has been prepared in a water bath; it is cooled to 70°C and added to the mixture. Finally, the ingredients (e) are added and the whole medium is thoroughly blended.

Next the warm medium is poured into sterilized 1-lb jam jars to a depth of about 0.5 inch and while still warm each jar is tipped and twisted so as to coat part of the interior. As soon as the medium is cool the jars are turned upside-down to prevent unnecessary contamination. They can be conveniently stored at about 12°C.

After the larvae are introduced the jars should be kept on their sides so that comparatively little frass falls on the medium. Whatman No. 1 filter paper can be used, at first, to close the jars, but when the larvae reach the fifth instar this should be replaced by a piece of "Terylene" gauze to allow increased ventilation. About 15 fifth-instar larvae, and still more younger larvae, can be kept in a jar.

Getzin (1962) succeeded in rearing virus-free cabbage loopers on an artificial diet. Since this species is very prone to infection with a nuclear polyhedrosis the eggs were surface-sterilized with 0.05% solution of sodium hypochlorite, containing 0.02% Triton X-100, for 1 hour. The eggs are partially dechorionated by this treatment so it is necessary to keep them moist; they should be placed on moist filter paper and incubated at 25°C. Under these conditions the larvae were all virus-free, so that it would appear that infection must usually arise from external contamination of the egg by virus.

The diet used was a slight modification of one used for *Heliothis zea* (Boddie) by Vanderzant *et al.* (1962), and contained the following ingredients: 3 gm wheat germ, 3.5 gm vitamin-free casein, 3.5 gm sucrose, 0.5 gm sodium alginate, 1 gm Wesson's salts, 1 ml B-vitamins plus inositol, 0.3 gm cholesterol, 0.1 gm choline chloride, 0.5 gm ascorbic acid, 0.12 gm sorbic acid, 2.5 gm agar, 18 gm macerated cabbage tissue, and 67 ml water.

The ingredients were mixed, sterilized, and cooled to 60°C before the ascorbic acid was added. The medium was then dispensed into 100×20-mm petri dishes.

For studying a nuclear polyhedrosis of *Prodenia litura* (Fabr.), Harpaz and Ben Shaked (1964) used a medium consisting of 95 gm wheat bran, 0.5 gm dry yeast, and 30 ml water. Hatching and larval survival were poor under these conditions but improved somewhat when the bran was replaced by lucerne (alfalfa) meal (Brazzel *et al.*, 1961).

Two per cent KOH was found to be a more satisfactory sterilizer for the egg surface than 0.1% $HgCl_2$.

Silkworm larvae can be raised from egg to adult on a semi-synthetic diet containing mulberry-leaf powder. Larvae are fed on sliced media placed on filter paper in a petri dish; 20 newly hatched larvae can be used in each dish and the medium is replaced twice daily. Composition of the diet in percentages is as follows: potato starch, 15; mulberry-leaf powder, 8; glucose, 6.5; sucrose, 6.5; defatted soybean casein, 27; unrefined soybean oil, 4; β-sitosterol, 0.4; Wesson's salt mixture, 1; ascorbic acid, 1; cellulose powder, 30.6; vitamin mixture (see below); distilled water, ml per gm of dry weight, 1.5.

Amounts of vitamins added per 9 gm dry diet: 30 μg biotin, 0.3 mg calcium pantothenate, 3 mg choline chloride; 30 μg folic acid, 3 mg inositol, 0.3 mg niacin, 0.15 mg pyridoxine HCl, 0.15 mg riboflavin, and 0.15 mg thiamine HCl. All vitamins, including ascorbic acid, are added in a solution which includes a small amount of ethanol (Ito and Horie, 1962).

The mass-rearing of the larvae of 9 noctuid species on a simple artificial medium has been described by Shorey and Hale (1965). Each of the diet ingredients, with the exception of the agar, is blended with half the total amount of water. The agar is separately dissolved in the remaining water at 100°C. The agar solution is cooled to less than 70°C and is mixed with the other blended ingredients. Six-ounce waxed-paper cups with standard cardboard lids are used as larvae-rearing containers, approximately 2 oz of medium being poured into each container.

The composition of the medium (in grams for 120–150 cups) is as follows: soaked pinto beans, 2133 (before the dried beans are used for diet preparation, they are brought to a boil in the water in which they were soaked overnight); dried brewers' yeast, 320; ascorbic acid, 32; methyl p-hydroxybenzoate, 20; sorbic acid, 10; formaldehyde (40%), 20; agar, 128; water, 6400.

Using the synthetic medium described (see p. 183) Ignoffo (1964b) has devised a simple method of bioassay for testing the activity of suspensions of cabbage-looper polyhedrosis virus.

The synthetic diet, which reaches a temperature of 75°C after the addition of melted agar, is cooled in an ice bath and then held at a temperature of 35–38°C. One milliliter of a known polyhedra suspension is diluted with 4 ml of sterile distilled water and this suspension poured into a sterile 250-ml beaker. The volume of liquid in the beaker is brought to 100 ml with cooled medium. After a thorough mixing, the medium is poured into a 90 × 25-mm petri dish where it solidifies in a layer about 17 mm deep. Cylindrical plugs, averaging 3.1 gm, are cut from the solidified medium with a No. 12 cork borer (15-mm diameter) and placed in 1-ounce portion cups, one plug per cup. The cups are then transferred to 150×25-mm petri dishes (7 cups to a dish), infested with 3-day-old larvae (1 larva per cup), cooled, placed at 28°C, and observed daily.

To determine the activity of a polyhedra suspension, it must be standardized by hemocytometer counts to 25×10^8 polyhedra per milliliter. This concentration can be approximated by triturating 25 diseased, fifth-instar larvae with 100 ml of water. Four decimal dilutions of the original suspensions are made, using 9-ml dilution blanks of sterile distilled water; 1 ml each of decimal dilution and the original suspension are pipetted into tubes containing 4 ml of sterile distilled water and the contents of each tube blended with 95 ml of cooled medium.

Average mortality for concentrations of 0, 50, 100, 150, 250, 350, 500, and 1500 polyhedra per μl was 0.0, 7.3, 26.2, 32.9, 45.1, 58.5, 81.7, and 96.4% respectively.

F. Staining Methods for Optical Microscopy

The routine method for observing the presence of polyhedra is to make a smear on a microscope slide from the larva under investigation, fix by gentle heat over a Bunsen burner and stain with Giemsa's solution. Nuclear polyhedra remain unstained against a stained background.

In an early study of a polyhedrosis of larvae of *Sphinx ligustri* (Linn.), it was observed that in a smear stained with Giemsa's solution, some polyhedra stained readily by comparison with others which remained unstained (Smith *et al.*, 1953). It was subsequently shown that those polyhedra which took the stain were cytoplasmic, as compared with the nonstaining nuclear polyhedra.

This staining technique is useful for the diagnosis of cytoplasmic polyhedra, but it is important not to apply too much heat when fixing or drying the polyhedra over the gas burner. It is usually necessary to pretreat nuclear polyhedra with weak acid before they will take a stain. After pretreatment with normal hydrochloric acid, nuclear polyhedra stain intensely with bromophenol blue, thus enabling their development to be followed (Xeros, 1953).

A method for the rapid staining of nuclear polyhedra in lepidopterous and hymenopterous hosts has been described by Smirnoff (1961). Immature polyhedra in the nuclei of the hemolymph cells are easily differentiated by staining the smears with Ziehl fuchsin or with Giemsa stains, but completely formed or purified polyhedra are much more resistant and can be stained only by amidoschwartz stain (Buffalo black) after treatment with a saturated solution of picric acid in water.

Another method of staining polyhedra *in situ* is as follows. After picric acid fixation and preparation with hemalum, aniline methyl blue with acid fuchsin is applied warm and differentiated with methanyl yellow, the latter stain having no effect on the polyhedra.

By this method polyhedra can be recognized at the beginning of their development, and can be distinguished from similar-appearing bodies of nonviral origin.

This technique has been used for the diagnosis of different viroses in Lepidoptera, the nuclear type in fat and hypodermic tissues, the cytoplasmic type in intestinal cells, and of nuclear polyhedra in the larvae of *Tipula paludosa* (Diptera) (Vago and Amargier, 1963).

Gershenson (1960b) has developed a method of contrast-staining for nuclear polyhedra with the idea of delineating the edges of the polyhedra. By this means it is possible to determine the exact crystalline shape of the inclusions. He suggests the use of a staining method whereby only the angles and edges of the polyhedra are stained, the bulk of the crystal remaining unstained. The method is first to mordant the polyhedra with 5% aqueous solution of phenol and then stain with carbolfuchsin. Penetration of the stain always starts from the edges and angles; by regulating the duration of the mordant treatment and of the staining, the stain can be confined to those regions. By the use of this method it is possible to determine the crystallographic topography in difficult cases.

The fluorescent-antibody technique has been adapted to demonstrate the presence of nuclear polyhedrosis in the silkworm. Fluorescence was observed in the cytoplasm before crystallization of the polyhedra, but fluorescing polyhedra were found only in the nuclei (Krywienczyk, 1963).

Huger (1961) has developed a staining method for delineating the capsules of granuloses based on the acid pretreatment used in the study of other inclusion-body diseases. His method is briefly as follows: infected tissues are fixed in Bouin-Duboscq-Brasil fluid, embedded in wax and sectioned in the usual manner. After dewaxing, the sections should be brought through descending dilutions of alcohol to distilled water and then transferred to 50% acetic acid at room temperature for 5 minutes. The sections are then rinsed in distilled water and mordanted in 2.5% iron alum for 2 hours; this is followed by further rinsing in distilled water and immersion in Heidenhain's iron hematoxylin for 5 hours. Sections should be differentiated carefully in 2.5% iron alum solution until the black nuclei and characteristic deep-black network of diseased cells become clearly outlined. Wash in running tap water and counterstain in an aqueous solution of 0.5% erythrosin for 2 minutes. Dehydrate, clear in xylol and mount in synthetic resin.

By this method the capsules are stained an intense bright red and are clearly contrasted against the pale gray-violet-colored cytoplasm.

Hamm (1966) has used a modified azan staining technique for inclusion-body viruses. When this simplified azan technique was applied to acid-treated sections of lepidopterous larvae infected with a nuclear polyhedrosis virus, the polyhedra stained bright red.

The following details of the method are given:

STAINS

Solution I. Dissolve 0.1 gm azocarmine G in 100 ml distilled water and boil the solution for 5 minutes. Allow to cool and add 2 ml glacial acetic acid. Filter before use.

Solution II. Dissolve in 100 ml distilled water, 1.0 gm phosphotungstic acid (PTA), 0.1 gm aniline blue (water-soluble), 0.5 gm orange G., and 0.2 gm fast green F.C.F.

STAINING PROCEDURE

 1. Sections from toluene via alcohols to water
 2. 50% acetic acid, 5 minutes
 3. Distilled water rinse, 5 seconds
 4. Azocarmine (solution I), 15 minutes
 5. Distilled water rinse, 5 seconds
 6. Aniline, 1% in 95% alcohol, 30 seconds
 7. Distilled water rinse, 5 seconds (change often)
 8. Counterstain (solution II), 15 minutes
 9. 50% alcohol, 10 seconds
10. Absolute alcohol, 2 changes, 30 seconds each
11. Toluene, 2 changes
12. Mount in neutral, synthetic mounting medium

RESULTS

Virus inclusion bodies, red
Epicuticle, red
Endocuticle, blue
Muscle, light-blue to blue-green
Epidermal cells, yellowish-green
Fat-body cells, yellowish-green with darker green nuclei
Nerve tissue, light-blue
Silk gland, green; contents, red or blue
Midgut-epithelium, green and blue
This technique has been used successfully on lepidopterous larvae infected with nuclear polyhedroses, cytoplasmic polyhedroses, and granuloses.

REFERENCES

Aizawa, K. (1961). Change in the shape of silkworm polyhedra by means of passage through chick embryo. *Entomophaga* 6, 197–201.

Aizawa, K. (1962). Infection of the greater wax moth, *Galleria mellonella* Linn., with the nuclear polyhedrosis virus of the silkworm, *Bombyx mori* Linn. *J. Insect Pathol.* 4, 122–127.

Aizawa, K. (1963). The nature of infections caused by nuclear-polyhedrosis viruses. *In* "Insect Pathology" (E. A. Steinhaus, ed.), Vol. 1, p. 381. Academic Press, New York.

Aizawa, K., Furuta, Y., and Nakamura, K. (1961). Selection of a resistant strain to virus induction in the silkworm, *Bombyx mori*. *Nippon Sanshigaku Zasshi* 30, 405–412 (English summary).

Aruga, H., Yoshitake, N., Watanabe, H., Hukuhara, T., Nagashima, E., and Kawai, T. (1961). Further studies on polyhedroses of some lepidoptera. *Japan. J. Appl. Entomol. Zool.* 5, 141–144 (in Japanese with English summary).

Aruga, H., Hukuhara, T., Fukuda, S., and Hashimoto, Y. (1963a). Interference between cytoplasmic polyhedrosis viruses of the silkworm, *Bombyx mori* (Linn.), and of the pine caterpillar, *Dendrolimus spectabilis* (Butler). *J. Insect Pathol.* 5, 415–421.

Aruga, H., Yoshitake, N., and Watanabe, H. (1963b). Interference between cytoplasmic polyhedrosis viruses in *Bombyx mori* (L). *J. Insect Pathol.* 5, 1–10.

Bawden, F. C. (1958). Reversible changes in strains of tobacco mosaic virus from leguminous plants. *J. Gen. Microbiol.* 18, 751–756.

Beckman, H. F., Bruckart, S. M., and Reiser, R. (1953). Laboratory culture of the pink bollworm on chemically defined media. *J. Econ. Entomol.* 46, 627.

Bird, F. T. (1955). Virus diseases of sawflies. *Can. Entomologist* 87, 124–127.

Bird, F. T. (1959). Polyhedrosis and granulosis viruses causing single and double infections in the spruce budworm *Choristoneura fumiferana* Clemens. *J. Insect Pathol.* 1, 406–430.

Bird, F. T., and Elgee, D. E. (1957). A virus disease and introduced parasites as factors controlling the European spruce sawfly, *Diprion hercyniae* (Htg.) in central New Brunswick. *Can. Entomologist* **89**, 371–378.

Brazzel, J. R., Chambers, H., and Hamman, P. J. (1961). A laboratory rearing method and dosage-mortality data on the bollworm, *Heliothis zea. J. Econ. Entomol.* **54**, 949–952.

David, W. A. L. (1957). Breeding *Pieris brassicae* L., and *Apanteles glomeratus* L. as experimental insects. *Z. Pflanzenkrankh. Pflanzenschutz* **64**, 572–577.

David, W. A. L., and Gardiner, B. O. C. (1952). Laboratory breeding of *Pieris brassicae* L., and *Apanteles glomeratus* L. *Proc. Roy. Entomol. Soc. London* **A27**, 54.

David, W. A. L., and Gardiner, B. O. C. (1960). A *Pieris brassicae* (Linn.) culture resistant to a granulosis. *J. Insect Pathol.* **2**, 106–114.

David, W. A. L., and Gardiner, B. O. C. (1965a). Rearing *Pieris brassicae* L. larvae on a semisynthetic diet. *Nature* **207**, 882–883.

David, W. A. L., and Gardiner, B. O. C. (1965b). Resistance of *Pieris brassicae* (Linn.) to granulosis virus and the virulence of the virus from different host races. *J. Invert. Pathol.* **7**, 285–290.

David, W. A. L., and Gardiner, B. O. C. (1965c). The incidence of granulosis deaths in susceptible and resistant *Pieris brassicae* (Linn.) larvae following changes of population density, food and temperature. *J. Invert. Pathol.* **7**, 345–355.

Day, M. F., and Mercer, E. H. (1964). Properties of an iridescent virus from the beetle, *Sericesthis pruinosa. Australian J. Biol. Sci.* **17**, 892–902.

Gershenson, S. M. (1960a). A study on a mutant strain of a nuclear polyhedral virus from the oak silkworm. *Probl. Virol. (USSR) (English Transl.)* **6**, 720–725.

Gershenson, S. M. (1960b). The geometrical shape of the intranuclear inclusions in insect polyhedroses. *Probl. Virol. (USSR) (English Transl.)* **5**, 109–113.

Getzin, L. W. (1962). Mass-rearing of virus-free cabbage loopers on an artificial diet. *J. Insect Pathol.* **4**, 486.

Glaser, R. W. (1915). Wilt of gipsy moth caterpillars. *J. Agr. Res.* **4**, 101–128.

Hamm, J. J. (1966). A modified azan staining technique for inclusion-body viruses. *J. Invert. Pathol.* **8**, 125–126.

Harpaz, I., and Ben Shaked, Y. (1964). Generation-to-generation transmission of the nuclear polyhedrosis virus of *Prodenia litura* (Fabr.). *J. Insect Pathol.* **6**, 127–130.

Huger, A. (1961). Methods for staining capsular virus inclusion bodies typical of granuloses of insects. *J. Insect Pathol.* **3**, 338–341.

Ignoffo, C. M. (1963). A successful technique for mass-rearing cabbage loopers on a semisynthetic diet. *Ann. Entomol. Soc. Am.* **56**, 178–182.

Ignoffo, C. M. (1964a). Production and virulence of a nuclear-polyhedrosis virus from larvae of *Trichoplusia ni* (Hüton) reared on a semisynthetic diet. *J. Insect Pathol.* **6**, 318–326.

Ignoffo, C. M. (1964b). Bioassay technique and pathogenicity of a nuclear polyhedrosis virus of the cabbage looper, *Trichoplusia ni* (Hüber.). *J. Insect Pathol.* **6**, 237–245.

Ishikawa, Y., and Asayama, T. (1961). Studies on the relation between the polyhedroses of the wild insects and the silkworm, *Bombyx mori* L. II. On the double infection of the nuclear polyhedroses in silkworm larvae. *Nippon Sanshigaku Zasshi* **30**, 201–205 (English summary).

Ito, T., and Horie, J. (1962). Nutrition of the silkworm, *Bombyx mori*, VII. An aseptic culture of larvae on semisynthetic diets. *J. Insect Physiol.* **8**, 569–578.

Kassanis, B. (1963). Interactions of viruses in plants. *Advan. Virus Res.* **10**, 219–253.

Krywienczyk, J. (1963). Demonstration of nuclear polyhedrosis in *Bombyx mori* L. by fluorescent antibody technique. *J. Insect Pathol.* **5**, 309–317.

Krywienczyk, J., and Bergold, G. H. (1960a). Serological relationships of viruses from some lepidopterous and hymenopterous insects. *Virology* **10**, 308–315.

Krywienczyk, J., and Bergold, G. H. (1960b). Antigenicity of insect virus membranes. *Virology* **10**, 549–550.

Krywienczyk, J., and Bergold, G. H. (1960c). Serological relationships between inclusion body proteins of some lepidoptera and hymenoptera. *J. Immunol.* **84**, 404–408.

Krywienczyk, J., MacGregor, D. R., and Bergold, G. H. (1958). Serological relationship of viruses from some lepidopterous insects. *Virology* **5**, 476–480.

Martignoni, M. E. (1964). Progressive nucleopolyhedrosis in adults of *Peridroma saucia* (Hübner). *J. Insect Pathol.* **6**, 368–372.

Martignoni, M. E., and Schmid, P. (1961). Studies on the resistance to virus infections in natural populations of lepidoptera. *J. Insect Pathol.* **3**, 62–74.

Neilson, M. M. (1965). Effects of a cytoplasmic polyhedrosis on adult lepidoptera. *J. Invert. Pathol.* **7**, 306–314.

Ossowski, L. L. J. (1958). Occurrence of strains of the nuclear polyhedral virus of the wattle bagworm. *Nature* **181**, 648.

Ossowski, L. L. J. (1960). Variation in virulence of a wattle bagworm virus. *J. Insect Pathol.* **2**, 35–43.

Rivers, C. F. (1959). Virus resistance in larvae of *Pieris brassicae* L. *Trans. 1st Intern. Conf. Insect Pathol. & Biol. Control, Prague, 1958* pp. 205–210.

Shorey, H. H., and Hale, R. L. (1965). Mass-rearing of the larvae of nine noctuid species on a simple artificial medium. *J. Econ. Entomol.* **58**, 522–524.

Sidor, Č. (1959). Susceptibility of larvae of the large white butterfly (*Pieris brassicae* L.) to two virus diseases. *Ann. Appl. Biol.* **47**, 109–113.

Smirnoff, W. A. (1961). Rapid staining of polyhedra from lepidopterous and hymenopterous hosts. *J. Insect Pathol.* **3**, 218–220.

Smith, K. M. (1963). The cytoplasmic virus diseases. *In* "Insect Pathology" (E. A. Steinhaus, ed.), Vol. 1, p. 457. Academic Press, New York.

Smith, K. M., and Xeros, N. (1953a). Studies on the cross-transmission of polyhedral viruses: Experiments with a new virus from *Pyrameis cardui*, the painted lady butterfly. *Parasitology* **43**, 178–185.

Smith, K. M., and Xeros, N. (1953b). Cross-inoculation studies with polyhedral viruses, *Symp. Interactions Viruses & Cells, Rome, 1953* pp. 81–96.

Smith, K. M., Wyckoff, R. W. G., and Xeros, N. (1953). Polyhedral virus diseases affecting the larvae of the privet hawk moth (*Sphinx ligustri*). *Parasitology* **42**, 287–289.

Smith, K. M., Hills, G. J., and Rivers, C. F. (1961). Studies on the cross-inoculation of the *Tipula* iridescent virus. *Virology* 13, 233–241.

Stairs, G. R. (1964). Selection of a strain of insect granulosis virus producing only cubic inclusion bodies. *Virology* 24, 520–521.

Steinhaus, E. A. (1957). New records of insect virus diseases. *Hilgardia* 26, 417–430.

Stephens, J. M. (1963). Immunity in insects. *In* "Insect Pathology" (E. A. Steinhaus, ed.), Vol. 1, p. 273. Academic Press, New York.

Tanada, Y. (1953). Description and characteristics of a granulosis virus of the imported cabbage worm. *Proc. Hawaiian Entomol. Soc.* 15, 235–260.

Tanada, Y. (1959). Synergism between two viruses of the armyworm *Pseudaletia unipuncta* Haworth (Lepidoptera, Noctuidae). *J. Insect Pathol.* 1, 215–231.

Tanada, Y., and Chang, G. Y. (1964). Interactions of two cytoplasmic polyhedrosis viruses in three insect species. *J. Insect Pathol.* 6, 500–516.

Tanada, Y., and Tanabe, A. M. (1965). Resistance of *Galleria mellonella* (Linn.) to the *Tipula* iridescent virus at high temperature. *J. Invert. Pathol.* 7, 184–188.

Thompson, C. G. (1959). Thermal inhibition of certain polyhedrosis virus diseases. *J. Insect Pathol.* 1, 189–190.

Vago, C. (1959). On the pathogenesis of simultaneous virus infections in insects. *J. Insect Pathol.* 1, 75–79.

Vago, C., and Amargier, A. (1963). Coloration histologique pour la différenciation des corps d'inclusion polédriques de virus d'insectes. *Ann. Epiphyties* 14, 269–274.

Vanderzant, E. S., and Reiser, R. (1956a). Aseptic rearing of the pink bollworm on synthetic media. *J. Econ. Entomol.* 49, 7–10.

Vanderzant, E. S., and Reiser, R. (1956b). Studies of the nutrition of the pink bollworm using purified casein media. *J. Econ. Entomol.* 49, 454–458.

Vanderzant, E. S., Richardson, C. D., and Fort, S. W. (1962). Rearing of the bollworm on artificial diet. *J. Econ. Entomol.* 55, 140.

Waterson, A. P. (1965), *in litt.*

Yamafuji, K., Omura, H., and Otomo, N. (1958). An immunological investigation of viral polyhedrosis. *Enzymologia* 19, 175–179.

Xeros, N. (1953). Origin and fate of the virus bundles in nuclear polyhedroses. *Nature* 172, 548–549.

CHAPTER XI

Plant Virus-Insect Vector Relationships

Introduction

It is not considered inappropriate, in a book on insect virology, to include a chapter which describes some of the relationships of insects with viruses of another kind. Plant viruses are more dependent upon insects and related arthropods for their dissemination in the field than any other type of pathogen. With some plant viruses, especially those which have leafhopper (Jassidae) vectors, the relationship between virus and vector is so close as to raise a doubt as to whether we are dealing with a plant virus or an insect virus.

That some plant viruses do actually multiply in the insects which transmit them, and may even cause disease in their vectors, is not merely an example of the versatility of viruses. It opens up a new field of investigation; in Chapter IX an account is given of the newly developing technique of insect tissue culture. In the not too distant future it should be possible to follow the development in cell culture, not only of the insect viruses but also of those plant viruses which multiply in insects.

In this chapter the reader can follow all gradations in the insect-plant virus relationship from the purely casual and mechanical contamination of the jaws to the close biological relationship to which we have briefly referred above.

It is not possible in a single chapter to deal adequately with all the minutiae of the relationship between plant viruses and their insect vectors but it is hoped to give a fairly comprehensive account of the main factors in these relationships.

A. Mechanical Vectors

Mechanical transmission of plant viruses in this context refers to a mode of transfer in which there is no evidence of any biological relationship between virus and vector. The question as to whether some aphid-borne viruses are mechanically transmitted is discussed in the section dealing with aphid-virus relationships and, with one exception, this section is concerned with virus transmission by insects with biting mouthparts. It is, of course, fairly obvious that any virus which is transmitted mechanically by an insect is also easily sap-transmissible; the four viruses mentioned here all fall into that category.

Perhaps the best example of a purely passive transfer of a plant virus is given by that of tobacco mosaic (TMV); Walters (1952) showed that the large grasshopper, *Melanoplus existientialis*, would carry the virus on its jaws and infect a healthy tobacco plant after first feeding on a mosaic-infected plant. The writer has also observed a certain amount of virus transfer on the feet of grasshoppers which have been allowed to wander, without necessarily feeding, on a mosaic-diseased tobacco plant and then transferred to plants of *Nicotiana glutinosa*. This plant reacts to infection by TMV with the production of localized foci of infection, and a few of these local lesions developed where the insects had placed their feet. Walters also demonstrated that the grasshopper would transmit potato virus X (PVX) by contamination of its jaws. Some experiments with the same virus were carried out by the writer using the potato flea beetle (*Psylliodes* sp.) as a possible vector; the results of these experiments were all negative. Schmutterer (1961) showed that TMV can be carried for a short period only on the mouthparts, but not the legs, of two beetles, *Tettigonia viridissima* and *T. cantans,* and of a caterpillar, *Barathra brassicae.* The virus appeared to be inactivated in less than an hour by the saliva, and cleaning the mouthparts of *T. viridissima* with KCN increased the infection rate. Heating to 65°C reversed the inactivation of TMV by salivary gland or intestine homogenate of *T. viridissima.* Exposure to higher temperatures increased the strong virus inactivation by the blood which, however, still contained traces of active TMV after long feeding periods. Schmutterer considers that virus inactivation is probably due to enzymes such as amylase. It is probable that the most important vector, in the field, of TMV and potato virus X is man himself by means of the contamination of his hands and implements.

Two good examples of the purely passive transfer of animal viruses by insects are the spread of fowl-pox virus on the proboscis of a bloodsucking fly and of the rabbit myxoma virus on the mouthparts of the mosquito, called by Fenner the "flying pin." We can find a parallel to this in the plant viruses; Costa *et al.* (1958) demonstrated the transmission of the virus of sowbane mosaic (SbMV) on the ovipositor of a fly (*Liriomyza langei* Frick) during the process of depositing its eggs in the epidermis of leaves in which the larvae subsequently live as "leaf miners."

The methods of mechanical transmission dealt with, so far, have been concerned only with external contamination; there are, however, other examples of mechanical transmission in which more than this is involved. These are viruses of which the natural vector is a beetle, an insect with biting mouthparts; a good example is the virus of turnip yellow mosaic (TYM), which is spread in the field by one or more species of flea beetle (*Phyllotreta* sp.). In the discussion on the transmission of plant viruses by aphids and other hemipterous insects it will be shown that the saliva plays an important part in that process; it is interesting therefore to find that the absence of saliva in biting insects is the essential fact in virus transmission. Although the turnip flea beetle is the natural vector of TYM virus it can be transmitted by any type of beetle which will feed on the necessary susceptible plants. The explanation seems to be as follows. Since beetles have no salivary glands it is necessary for them to regurgitate part of the contents of the foregut while eating; this apparently helps in the process of digestion. Regurgitation brings into contact with the leaf any infective tissue previously eaten and this, during the process of mastication, is inoculated into a healthy susceptible plant. Transmission of this type is not confined to beetles, so far as TYM virus is concerned, and it can be spread by grasshoppers and earwigs (Forficulidae) which feed in a similar manner. The virus, however, is not spread by caterpillars, which are also biting insects, but which do not regurgitate while feeding. Transmitting beetles or their larvae can retain infective power for several days without further access to a source of virus; this is thought to be due merely to the length of time infective tissue remains in the foregut. When this is digested the insect must feed once more on an infected plant before it can again transmit the virus.

B. The Vector Relationships of Tobacco Mosaic Virus

It has long been a puzzle to plant virologists that tobacco mosaic virus (TMV), the most infectious plant virus known, has no efficient vector other than man himself. We have already considered in Section A the purely mechanical transmission of TMV by biting insects, such as grasshoppers and beetles (Walters, 1952), so that aspect of the subject will not be referred to again, except incidentally.

The literature contains many references to apparently successful transmission of TMV by different species of insects, mainly belonging to the Hemiptera, but attempts to repeat these transmissions have invariably failed.

The earliest record seems to be the report of Allard (1914, 1917) that a virus, which he believed to be TMV, was transmitted by an aphid. It is possible, in such early days, that Allard mistook for TMV a strain of cucumber mosaic virus (CMV) which, of course, is easily transmitted by aphids; there are several strains of CMV which, from the symptoms produced on the tobacco plant, might be confused with TMV.

It is more difficult, however, to explain the successful results of Hoggan (1931, 1934), who carried out an extensive investigation on the transmission of TMV by aphids. According to this work, 4 species of aphids transmitted the virus from tomato to tobacco and various other solanaceous hosts, but not, apparently, from tobacco to tobacco. The aphids used were, *Aulacorthum solani* (Kalt.) (= *Myzus pseudosolani* Theob.), *Macrosiphum euphorbiae* (Thomas) [= *M. solanifolii* (Ashmead)], *A. circumflexus* (Buckt) [= *Myzus circumflexus* (Buckt)], and *Myzus persicae* (Sulz). Of these, *M. persicae* was the least efficient vector. Hoggan used large numbers of aphids, about 300 per plant; they were allowed to feed on the infected plants for 2–5 days and were then transferred on detached leaves, placed on pieces of paper near the plants to be colonized. In another experiment, Hoggan (1934) transferred the aphids from the TMV-infected plants to hosts which react only with local lesions. One lesion was produced by *each* 129 *A. solani,* 140 *M. euphorbiae,* and 812 *M. persicae.*

There are several other recorded instances of apparently successful transmission of TMV by aphids and other sap-sucking insects. Cornuet and Morand (1960a,b) describe a strain of this virus which affects *Fragaria vesca* L. In their experiments Cornuet and Morand claim to have transmitted this strain of TMV, by means of the aphid *Pentatrichopus fraegaefolii* Cock, not only to strawberry plants but also from tomato and tobacco to tobacco.

Early claims to have transmitted TMV by mealybugs were made by Olitsky (1925) and Elmer (1925), using *Pseudococcus citri* (Risso) and *P. maritimus* Ehr., respectively. Newton (1953), using *P. citri,* obtained 100% transmission with 4 insects on each of 5 tobacco plants, and on a local lesion host 1–3 lesions developed after the feeding of 10 mealybugs.

Orlob (1963), who has made a reappraisal of the transmission of TMV by insects, repeated all these apparently successful experiments with uniformly negative results. In Table XI is reproduced Orlob's table in which the results of previous authors' work are compared with those of his experiments.

Brčák (1959) also was unable to transmit TMV with 300 *P. maritimus* transferred to 71 test plants or with 500 whiteflies (Aleyrodidae).

It has been suggested that aphids cannot transmit TMV because they do not imbibe the virus. Sukhov (1944) thought that the salivary sheath surrounding the aphid puncture in the tissue acted as a filter and prevented the uptake of virus. Van Soest and de Meesters-Manger Cats (1956) were unable to detect the presence of TMV in the droplets exuding through stylets, cut off but left *in situ*, by means of the electron microscope or by serological means, and they also concluded that the aphid *Myzus persicae* did not take up TMV from an infected plant. However, it has now been proved that both aphids and leafhoppers do in fact imbibe this virus. Kikumoto and Matsui (1962) demonstrated the presence of TMV rods in the midguts of aphids which had fed on infected plants (see also Ossiannilsson, 1958).

Several workers have shown that TMV can pass through the gut of biting insects and can be recovered from the feces (Brčák, 1957; Walters, 1952). The writer (Smith, 1941) was unable to transmit TMV by means of several species of caterpillars but viable virus was easily recoverable from the feces. Orlob (1963) was also able to recover virus from the feces of the potato flea beetle, *Epitrix cucumeris* Harr.; in addition he found this beetle could act as an inefficient mechanical vector, the virus only being transmitted if the insect scraped the epidermis of the leaf without ingesting the tissue.

What then can be the reason for the nontransmission of TMV by hemipterous insects, especially aphids? In the next section we shall discuss a type of mechanical transmission by aphids in which the virus adheres to the distal end of the stylets. Why is the very infectious TMV not transmitted in a similar manner? It cannot be a question of the shape of the virus particle; *Hyoscyamus* mosaic virus and potato virus Y (PVY), both rod shaped, are easily aphid-transmitted but TMV and potato virus X (PVX) also with rod-shaped particles, are not so transmitted.

Orlob (1963) has carried out a number of experiments to investigate

TABLE XI

RESULTS OF TRANSMISSION OF TOBACCO MOSAIC VIRUS BY SUCKING INSECTS OBTAINED BY VARIOUS AUTHORS, COMPARED WITH RESULTS OF PRESENT WORK[a]

Experiment no.	Author	Virus source	Test plant	Vector	No. of insects per test plant or leaf	Results	
						Author	Present work
1	Hoggan	Lycopersicon esculentum	Nicotiana tabacum	Macrosiphum euphorbiae	300	30/50[b]	0/12[b]
				Aulacorthum solani	300	39/50[b]	0/12[b]
2	Hoggan	L. pimpinellifolium	N. glutinosa	M. euphorbiae	?	27/3775[c]	0/785[c]
3	Cornuet and Morand	Fragaria vesca	F. vesca	Pentatrichopus fragaefolii	15	+[d](4/4[b]?)	0/20[b]
4	Cornuet and Morand	N. tabacum	N. tabacum	P. fragaefolii	20	+(1/2[b]?)	0/20[b]
5	Olitsky	N. tabacum	N. tabacum	Pseudococcus citri	5	3/5[b]	0/8[b]
6	Olitsky	N. tabacum	N. glutinosa	P. citri	10	1-3/10[c]	0/20[c]
7	Newton	N. tabacum	N. tabacum	P. citri	4	+5/5[b]	0/8[b]

[a] After Orlob (1963).
[b] Number of test plants infected per number colonized.
[c] Number of local lesions per total number of insects used.
[d] + = Positive result.

the relationships between the aphid's stylets and TMV. Wetting the stylets of aphids or other sap-sucking insects and the mouthparts of the potato flea beetle with TMV or with TMV nucleic acid failed to make any insects viruliferous with the exception of the last-named. Virus could be recovered from the bared stylets but not after the labium had extended down over them.

It has now been proved that TMV is imbibed by aphids and other sucking insects and it can be demonstrated inside the midgut of aphids (Kikumoto and Matsui, 1962). What then prevents the virus from passing out of the gut and reaching the saliva and thus being returned to the plant? Although the presence of the two rod-shaped viruses of *Hyoscyamus* mosaic and PVY, which are aphid transmitted, has not been demonstrated in the midgut of aphids, they can be legitimately assumed to be there. However, this is not necessary for the transmission of these two viruses, since they adhere to the stylets and are transmitted as soon as the stylets enter a susceptible cell. We have, then, apparently comparable aphid relationships of TMV, *Hyoscyamus* mosaic virus, and PVY, namely, presence of the virus in the gut and adherence to the stylets, but there is still some factor preventing the transmission of TMV. Can the explanation be a purely quantitative one? Orlob has shown that there is very little virus left on the stylets once the labium has extended over them. Yet this must apply in an even greater degree to the other two viruses which are aphid transmitted. In the present state of our knowledge of the subject we can only assume that aphids do not transmit TMV because not enough virus adheres to the stylets to make an infective dose. There is some evidence in support of this in the fact that some biting insects easily transmit the virus provided there is gross contamination of their jaws.

A Japanese worker Nishi (1962), quoted by Maramorosch and Jensen (1963), suggests that the reason aphids cannot transmit TMV is the toxicity of the insect's saliva to the virus. Homogenates of the salivary sheaths left by the aphid after feeding (Fig. 44) were found to have an inhibitory substance which acted directly on the virus without affecting the test plants. When aphids were fed through membranes on sugar solutions, and the solutions, now containing aphids' saliva, were mixed with TMV and used for inoculum, a similar inhibition was observed.

Electron micrographs of the inhibited TMV particles indicated that the inhibitor did not combine with TMV, nor destroy the virus, but, instead, covered the particles with a granular membrane.

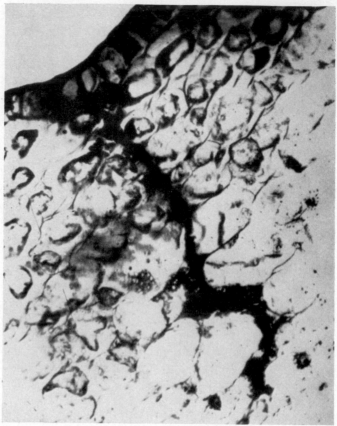

Fig. 44. Section through a potato leaf showing stylet track of the aphid *Myzus persicae* Sulz. Note that the stylets have penetrated intercellularly. Magnification, × 1200.

C. The Problem of Aphid-Virus Relationships

In this section we can visualize the gradual development from the purely passive transfer of virus to more intimate relationships between insect vector and plant virus.

The problem of aphid-virus relationships is no simple one, and, in spite of many ingenious experiments and much theory, there is still considerable confusion of thought. Perhaps the best way to present the problem will be to state the many facts which have been discovered and then to discuss the various explanations which have been offered.

The pioneers in this field were Watson (1936, 1938, 1946) and Watson and Roberts (1939). Using the virus of *Hyoscyamus* mosaic, they confirmed earlier observations which had been made on the aphid transmission of cucumber mosaic virus and potato virus Y, namely, that some viruses were rapidly lost by the aphid vector. Aphid-transmitted viruses were divided into two categories, the *nonpersistent* (or external) which were usually lost by the aphid after feeding on one or two plants, and the *persistent* (or internal) which were retained by the aphid for long periods.

It was recognized, however, that there was no hard and fast dividing line and that these two categories tended to merge into each other.

Here we may digress for a moment to discuss the nomenclature of these different types of aphid-virus relationships. A third category was added, the *semipersistent* viruses, to include those cases in which retention of virus by the aphid is relatively brief but is, nevertheless, considerably longer than in the case of the nonpersistent viruses.

Just recently Kennedy *et al.* (1962) have introduced terms which are less empirical and which give some indication of the location of, and route followed by, the virus in the insect. For nonpersistent viruses the term *stylet-borne* is suggested; persistent viruses become *circulative* and those viruses which have a definite biological relationship with their vectors are known as *propagative*. This type of virus is more usually associated with leafhopper vectors, but, as we shall see later, there is evidence that they are also associated with aphids. That leaves the semipersistent viruses unaccounted for, and it seems a matter of choice as to which category they should be assigned; thus Maramorosch (1963) designates them as stylet-borne, and Sylvester (1962), for convenience, and for lack of more precise knowledge, includes them in the circulative viruses.

These terms suggested by Kennedy *et al.* seem likely to be adopted by plant virologists and they will be used throughout this review.

We shall now return to the discussion of these three types of aphid-virus relationships, giving the facts known about each and the various explanations put forward concerning them.

a. STYLET-BORNE VIRUSES

The following facts about the stylet-borne viruses have been elucidated by Watson and her co-workers.

1. A preliminary fasting period by the aphid greatly increases the capacity to transmit virus.

2. A short feed (acquisition feed) is more effective than a long period of feeding, and aphids making many penetrations retained infectivity longer than those making prolonged single penetrations.

3. Prolonged feeding reduces the effciency of virus transmission.

In discussing the various theories put forward in explanation of these facts, it must be borne in mind that there is another phenomenon in connection with the stylet-borne viruses: that is the question of aphid specificity. In other words, one species of aphid can transmit a particular virus while another apparently similar species cannot do so.

Some rather complicated, but not very precise, theories have been put forward by various workers to account for these phenomena, and to explain the question of aphid specificity. Watson and Roberts (1939) were the first to suggest an "inactivator-behavior theory" and this was followed by Day and Irzykiewicz (1954) who introduced a *mechanical-inactivator-behavior* hypothesis, which suggested contamination of the stylets together with selective inactivation of virus on the stylets by fasting labile components of the saliva. Differences in behavior of the different aphid species might help to account for selective transmission. As we shall see later, the behavior of the aphids themselves plays an important role in virus transmission.

Van der Want (1954) suggested a *mechanical-surface adherence* hypothesis; here emphasis is laid on surface differences of the stylets of various aphid species, so that specificity is tied to differential adsorption to, and elution of the virus from, the stylets. In this connection it should be mentioned that van Hoof (1957) found differences in structure of the stylets of various aphids. Low-power electron micrographs revealed the presence of different excrescences, pockets, or ridges to which virus might be expected to adhere.

Sylvester (1954) proposed a *mechanical-inactivator-compatibility* hypothesis in which specificity depended upon the compatibility of the combination of the virus, saliva, and host cell inoculated; the actual transmission being effected by virus carried in or on the stylets.

The idea of a purely mechanical process was reintroduced by Bradley (1952), and later (1959) he emphasized the role of behavior in at least some of the transmission phenomena. This was an important observation, and the relevance of aphid behavior and the light it throws on some of the rather peculiar facts stated above, will now be discussed. It was noticed that after a fasting period the aphid tended to make a number of very short penetrations of the leaf, mainly in the epidermis where many

of the stylet-borne viruses are located. Furthermore, saliva was not se-
creted and virus contamination occurred during these short penetrations.

Bradley also observed that many aphids when removed from a feeding position have the stylets protruding beyond the tip of the rostrum.
It is not possible for these thin and delicate stylets to pierce the leaf epidermis unsupported; the aphid must therefore wait until the rostrum is
again extended down over them. The behavior of the aphid then offers
two reasons why a preliminary starving period enhances virus transmission. First, it allows time for the extended stylets to be restored to their
proper position for leaf penetration; and second, it encourages numerous
short stabs into the epidermis which appear to offer optimum conditions
for transmission.

Another of the observations made by Watson was that the longer the
aphid fed on the diseased plant the less efficient it became as a vector.
This may be partly explained by the fact that the stylets penetrate, during prolonged feeding, into deeper tissue where the stylet-borne viruses
are not located. Sylvester (1962), who has made a careful study of the
feeding habits of aphids, has suggested what may be called an *infective
salivary* plug concept of transmission, as an additional reason for this
phenomenon. When a fasted aphid lowers its rostrum to the leaf epidermis, saliva begins to flow. The stylets begin to penetrate the cuticle and
the stylet movements mold the salivary liquid into a gelatinous tube
which eventually congeals to surround and support the flexible stylets
(Fig. 44). Sylvester noticed that during brief penetrations of 15 to 20
seconds the salivary sheath was incomplete, and examination of the stylets, after such a brief penetration, revealed a plug of saliva clinging to
the tips of the stylets which is brought into contact with the tip of the
labium when the latter ensheathes the stylets. Thus virus-contaminated
material will be smeared over the tip of the rostrum and the stylets and
inoculated into the cell at the next penetration. On the other hand, prolonged feeding tends to eliminate transmission as the stylets would have
to be drawn through the long close-fitting tube shown in Fig. 44 and this
would tend to remove any virus-contaminated material adhering to them.
It is relevant here to draw attention to some experiments by Bradley
(1956), who showed that penetration through a thin membrane,
stretched over the leaf surface, reduced the efficiency of both acquisition and inoculation probes.

To sum up, then, we can partially explain the transmission of the
stylet-borne viruses by mechanical adherence of the virus to about 15 mμ

of the distal end of the stylets. Most of the other phenomena associated with this type of virus transmission are explainable in terms of aphid behavior. The problem of aphid specificity, however, is still not satisfactorily solved.

b. CIRCULATIVE VIRUSES

As we have mentioned earlier, there is lack of precise knowledge concerning the semipersistent type of viruses. Some workers consider they are stylet-borne while others put them in the category of circulative viruses, in which they are included here for convenience. One factor in deciding whether a semipersistent virus belonged in the stylet-borne or circulative groups would be its retention through a molt. When molting occurs, the interior linings of the fore- and hindgut are shed along with the stylets and skin of the insect. Hence, any virus transmitted after a molt, provided the aphid had no access to a further source of virus, must have been present in the midgut. However, so far as the semipersistent virus of sugarbeet yellows is concerned, repeated attempts to test for retention of this virus after molting have been unsuccessful (Day, 1955; Sylvester, 1962). Some recent work by Nault *et al.* (1964) shows that the circulative virus of pea enation mosaic (PEMV) is retained by the aphid vector after molting. Virus was retained by 16 aphids through at least one molt; 11 of 16 aphids transmitted virus after 3 molts. Some aphids were injected with hemolymph from other aphids which had been reared on infected pea plants, and in 3 out of 4 experiments the injected aphids transmitted PEMV to healthy plants. Nymphs were more efficient than adults in transmitting the virus during short probes, and the fact that the virus can be spread by short probes suggests that, unlike most circulative aphid-borne viruses, it may not be necessary for PEMV to be inoculated into the phloem of the susceptible host. Although it cannot definitely be said that PEMV multiplies in the aphid vector, this work demonstrates a biological relationship of some kind between virus and vector.

The chief differences between the true stylet-borne (nonpersistent) and the semipersistent viruses are the longer periods of retention by the aphid of the semipersistent virus, and the facts that preliminary starving does not influence subsequent transmission, while prolonged initial feeding increases rather than decreases the likelihood of virus transmission.

With most of the circulative viruses (persistent) there is a delay in the development of the infective power within the insect. In other words, the aphid does not become capable of transmitting the virus until a certain period has elapsed after acquisition of the virus. This latent period varies considerably from a few hours to several days; those which have the longest latent period are strawberry virus 3 (Prentice and Woollcombe, 1951) and sow thistle yellow vein virus (Duffus, 1963). In the first case this varies from 10–19 days, and in the second 8–46 days according to the temperature.

Other circulative viruses are those of potato leaf roll (Smith, 1931a; Day, 1955), pea enation mosaic (Osborn, 1935; Chaudhuri, 1950; Simons, 1954), barley yellow dwarf (Toko and Bruehl, 1959) and radish yellows (Duffus, 1960). The two components causing the rosette complex disease of tobacco, the vein-distorting and mottle viruses, are also both circulative; the aphid vector *Myzus persicae* retains these for a period of 30 days and probably for the remainder of its life (Smith, 1946).

c. PROPAGATIVE VIRUSES

In the discussion on the relationships of plant viruses with leafhopper vectors convincing evidence that certain viruses do multiply in these insects will be presented. The situation as regards propagative viruses in aphids is less clear-cut, but there are two instances where the evidence for virus multiplication in the aphid is fairly convincing.

The first of these concerns the virus of potato leaf roll and the aphid vector *Myzus persicae*. It was with this virus that the serial inoculation technique, long used with leafhoppers, was first applied to aphids.

It was first shown by Smith (1931a) that the aphid *M. persicae* retained infectivity with the leaf roll virus, even after feeding on an immune plant, such as cabbage, for as long as 7 days. In 1955 Day also presented evidence which suggested the possibility that the potato leaf roll virus might multiply in the aphid, *M. persicae*. In the same year Heinze (1955) carried out experiments on the inoculation of aphids with plant extracts which opened the way to the subsequent work carried out by Stegwee and Ponsen (1958). Aphids were injected with aphid hemolymph and not with plant extracts which tended to be toxic. It was found that after injection with hemolymph from virus-bearing (viruliferous) aphids, about 50% of the injected aphids became infective after an average incubation period of 20 hours. When the hemolymph was diluted with saline the incubation period was extended to 7 or 10 days. Once infected with the potato leaf roll virus by injection the aphids retained their infectivity. Serial transmissions were then carried out as follows: Viruliferous aphids were colonized on a virus-immune plant such as Chinese cabbage for 7 days, hemolymph from these aphids was then injected into a series of known virus-free aphids. After 7 days the process was repeated, and this was carried out 15 times. At each passage the presence of the leaf roll virus was demonstrated by feeding the aphids on susceptible plants. It is calculated that if no multiplication had taken place the dilution would have reached 10^{-21}, while the actual dilution end point of the virus in hemolymph is 10^{-4}.

A method of isolating the potato leaf roll virus from the hemolymph of viruliferous aphids by means of gradient centrifugation has recently been described (Stegwee and Peters, 1961).

As we shall see later in discussing the biological relationship of some plant viruses with their leafhopper vectors, in some cases there is an adverse effect of the virus upon the insect vector. The question then arises: Is there any histological or metabolic disturbance caused by the leaf roll virus in the aphid? Opinion is divided on this question, although Schmidt (1959, 1960) found differences in the shape and size of the nuclei in the alimentary canal; in the viruliferous aphids the nuclei were smaller and of abnormal shape.

Ehrhardt (1960) studied the metabolic changes in the aphid *Myzus persicae* and observed a 30% decrease in oxygen consumption in viruliferous, as compared with virus-free, aphids.

A second case of the apparent multiplication of a plant virus in an aphid has been described by Duffus (1963). The virus in question, causing the yellow vein disease of sow thistle, *Sonchus oleraceus,* is transmitted by the aphid *Amphorophora lactucae* (L.). The latent period is very long and transmission is independent of the quantity of virus ingested. The shortest insect-latent period recorded was 8 days at 25°C; the longest latent period was 46 days at 5°C. From experiments carried out on the incubation and latent periods in both the plant and the insect, it seems that a certain virus concentration must be built up in the aphid vector before transmission occurs. Aphids retain infectivity throughout their lives. The sow thistle yellow vein virus (SYVV) is transmitted in essentially the same manner as those plant viruses which multiply in their leafhopper vectors (see Section D,d).

The only other plant virus known at present, with a similar long latent period in the aphid vector, i.e., 10–19 days, is that of strawberry crinkle (Prentice and Woollcombe, 1951).

d. Variation in Transmission of a Single Virus by a Single Aphid Species

As Rochow (1963) has pointed out in this review of aphid variation in transmission, there are three kinds of variation in an aphid species, among different *clones* or *strains*, among various *developmental* stages, and among *different forms* of one species. Stubbs (1955) working with the yellows virus of spinach, a virus of the circulatory type, found considerable variation in the ability of different cultures of *Myzus persicae* to transmit the virus.

Björling and Ossiannilsson (1958) studied the transmission by *Myzus persicae,* among other species, of two viruses, those of beet yellows and potato leaf roll, a circulative and a propagative virus, respectively.

They found that they could arrange the 85 strains of *M. persicae* into groups in which the percentage of transmission ranged from 10 to 80%, the majority falling into the 40–50% group. These differences were about the same for both viruses and the ability to transmit was not related to the host plant from which the aphids originally came.

Björling and Ossiannilsson made crosses among the different clones of the aphid and obtained some evidence of the genetic basis for the variation.

Rochow (1960a) has found considerable differences in the ability of another aphid (*Toxoptera graminum* Rondani) to transmit the virus of barley yellow dwarf, another circulative virus. He made collections of the aphid from Florida, Wisconsin, and Illinois, and found that those from Wisconsin and Illinois were fairly efficient vectors but aphids from Florida were virtually unable to transmit the virus. Rochow points out a slight anatomical difference in the beak tip of the inactive clones of aphid.

Somewhat similar clonal variation occurs in the transmission of the stylet-borne viruses. Simons (1954) showed that the cotton aphid (*Aphis gossypii* Glover) differed in its ability to transmit the virus of southern cucumber mosaic. He also found that these differences are not paralleled in the transmission by the same aphid of a different stylet-borne virus such as potato virus Y.

There are two records in the literature of differences in transmitting power according to the *developmental* stage of the aphid; both are concerned with circulative viruses. In one of these the young nymphs of *Macrosiphum geranicola* (Hille Ris Lambers) acquired the virus of filaree red leaf virus more readily than the adults and the minimum latent period was shorter in the nymphs (Anderson, 1951). The second case concerns the virus of pea enation mosaic of which the first-instar nymphs of the pea aphid (*Macrosiphum pisi* Kalt) are more efficient vectors and have a shorter latent period than the adults (Simons, 1954).

Variation in the ability of different *forms* of the same aphid species has been observed on several occasions, but there seems to be little correlation. For example, Paine and Legg (1953) were able to transmit the virus of hop mosaic by means of the spring winged form of the aphid *Phorodon humuli* (Schrank) but not by the wingless form. On the other hand, Orlob and Arny (1960) found the wingless oviparous forms and the fundatrices of *Rhopalosiphum fitchii* (Sanderson) to be capable of transmitting barley yellow dwarf virus, but not the other types including the winged forms.

Again, Orlob (1962) found that the winged migratory form of *Aphis nasturtii* (Kalt) could transmit potato virus Y, but not the oviparae and fundatrices which spend their entire life cycle on the winter host.

e. VARIATION IN TRANSMISSION OF A SINGLE VIRUS BY DIFFERENT APHID SPECIES

The question of why one aphid species can transmit a particular virus while a second aphid species is not able to do so has already been discussed in the section dealing with stylet-borne viruses. In this section another aspect of the situation is briefly considered in which the inability to transmit a particular virus by a particular aphid species appears to be due to a change in the virus itself. As regards stylet-borne viruses a good example is that of cucumber mosaic virus and the aphid vector *Myzus persicae*. Badami (1958), working with a strain of this virus, isolated from spinach, which was easily transmitted by *M. persicae*, found that the aphid lost its transmitting ability so far as this strain of the virus was concerned. The aphid, however, was still able to transmit other strains of cucumber mosaic virus and the original strain was still aphid-transmissible by other species such as *Aphis gossypii* and *Myzus ascalonicus*.

The loss of their insect vector relationships by viruses is no new thing. It seems to occur when a virus has been propagated mechanically or by vegetative reproduction of the plant host over periods of years. This appears to have happened with the virus of potato paracrinkle after years of virus propagation by the tubers. It has also been recorded with the virus of tomato spotted wilt and the thrips vector and with some leaf-hopper-borne viruses where there has been no contact between virus and vector for periods of years. In the case of cucumber mosaic virus and *Myzus persicae,* the case is slightly different because the loss of vector relationship is confined to one virus strain only and to only one out of several vectors.

Another example of variation in transmission is that of barley yellow dwarf virus, of the circulative type. There is a remarkable specialization among isolates of the virus and some of the aphids that transmit it. Five aphid species were shown to be vectors and others have been described. The original studies were made by Oswald and Houston (1953) and since then many papers have been published by other workers, notably Toko and Bruehl (1956, 1957, 1959), Slykhuis *et al.* (1959), Watson and Mulligan (1960), and Rochow (1959, 1960b, 1961). The whole subject has recently been reviewed by Rochow (1963) and much of this information is taken from his paper.

Direct comparisons of three aphid species and different isolates of barley yellow dwarf virus have shown the existence of at least three vector-specific strains of the virus. One strain (MGV) is transmitted efficiently by *Macrosiphum granarium* (Kirby) but not transmitted regularly by *Rhopalosiphum padi* (L.) or *R. maidis* (Fitch). A second strain (RPV) is transmitted efficiently by *R. padi* but not transmitted regularly by *M. granarium* or *R. maidis.* The third strain (RMV) is transmitted fairly efficiently by *R. maidis* but not transmitted regularly by *M. granarium* or *R. padi.*

In discussing the basis for this specificity Rochow suggests that the major basis for the variation lies in the virus rather than in the aphids. As to the mechanism controlling this specificity it might be centered in any one, or in a combination, of three main areas: acquisition of virus from the plant, physiology of virus in the aphid after acquisition, or inoculation of virus into plants by the aphid vector. The evidence suggests that failure to acquire virus from a source plant is not the major basis for the specificity. First the specificity persisted whether acquisition of the virus was by feeding or by injection (Mueller and Rochow, 1961). Second, it was shown by direct tests that the "nonvectors" actually acquired the virus (Rochow and Pang, 1961). Rochow concludes that the results of the various studies on vector specificity of barley yellow dwarf virus are perhaps best explained by assuming that the virus is extremely variable and that existing strains differ in the efficiency with which each aphid species can transmit them.

f. "Helper" Viruses and Aphid Transmission

The manner in which plant viruses react upon each other in the same hosts is a large and fascinating subject which has recently been discussed in a review by Kassanis (1963). One aspect of this behavior relates to the transmission of virus complexes by aphids. Selective transmission by the aphid is of various kinds; for example, when fed upon a potato plant containing the X and Y viruses the vector selects out virus Y because virus X is not aphid-transmitted (Smith, 1931b). Again from a complex of cabbage black ringspot and cauliflower mosaic viruses in the same host, the aphid M. persicae will transmit both viruses, whereas another aphid species will select one virus only (Kvíčala, 1945). There is, however, another related phenomenon which is little understood; that is the dependence of one virus upon another for aphid transmission. In other words, it is necessary for a second virus to be present in the plant with the first virus for the latter to be picked up, or at least transmitted by, the aphid.

The term "helper" virus is suggested for the second virus since the first is dependent upon it for its spread to other hosts.

The first example of this phenomenon was pointed out by Clinch et al. (1936), who stated that potato aucuba mosaic virus was only transmitted by aphids in the presence of potato virus A. They did not, however, present any experimental data, and the first study of this phenomenon was made by Smith (1946) on the virus complex in tobacco, known as "rosette." Two viruses are present in this disease, the vein-distorting and the mottle viruses; of these two only the mottle virus is sap-transmissible. Both viruses are of the circulative type and are retained by the aphid vector, Myzus persicae, for long periods, generally for the remainder of its life. From the rosette disease the aphid transmits both viruses together with great regularity. When the two viruses are separated, however, and exist alone in individual tobacco plants, only the vein-distorting virus is aphid-borne.

In experiments performed at intervals over a period of years the mottle virus has never been aphid-transmitted. Various explanations of this phenomenon have been put forward but there is little supporting evidence for any of them. It was thought that possibly the presence of the vein-distorting virus increased the concentration of the mottle virus to a point where it could be picked up by the insect, but dilution tests do not bear this out. Another possibility is that the vein-distorting virus adsorbs the mottle virus and so may be said literally to carry it. Examination of the leaf tissues of rosette plants, however, by ultrathin sections and electron microscopy has so far failed to reveal the viruses.

Kassanis (1961) has investigated further the problem of the aphid transmission of potato aucuba mosaic virus and potato viruses A and the closely allied virus Y; these are all stylet-borne viruses compared to the circulative-type viruses of the rosette complex. Kassanis found that not all the strains of potato aucuba mosaic virus behaved similarly; the ease with which they become aphid-transmissible varied and some strains did not become aphid-transmissible at all. Also two strains were transmitted in the presence of potato virus Y but not in the presence of potato virus A.

What appears to be a similar case of the "helper" virus has recently been reported by Watson (1962). She has worked on the yellow net virus of sugarbeet in England, and she found that the yellow net virus (YNV) was invariably accompanied in the sugarbeet by a second virus, yellow net mild yellows (YNMY) without which YNV could not be aphid transmitted. Watson refers to YNMY virus as a "carrier" virus and suggests that it may help YNV to invade sugarbeet or the two may combine to form an aphid-transmissible combination, one virus presumably being adsorbed by the other. However, Kassanis (1963) thinks there is not yet sufficient experimental evidence that this is another case of the "helper" virus. He points out that the interaction was suggested because YNV is rarely transmitted from *Nicotiana clevelandii*, which is immune to YNMY, but can easily be transmitted from sugarbeet plants in which it is always found associated with YNMY. It is rather surprising that the yellow net virus should always have been transmitted together with the mild virus. Kassanis suggests that a simpler explanation for the results is that the aphids very rarely transmit yellow net virus from *N. clevelandii*. He quotes an apparently similar situation in which dandelion yellow mosaic virus is transmitted by aphids with difficulty from dandelion to lettuce plants but readily from lettuce to lettuce plants (Kassanis, 1947). Furthermore the existence of YNMY has not been reported in the United States where YNV is aphid-transmitted (Sylvester, 1949).

g. Artificial Feeding and Virus Injection of Aphids

Much of our knowledge on the relationships of plant viruses and leafhoppers, which are discussed in the next section, has arisen from the artificial feeding and injection methods that have been developed. Similar techniques applicable to aphids have been slower in development because of the more fragile nature of these insects and the presence in the body of large numbers of developing young. However, as we have previously pointed out, these two techniques are now being used successfully, largely because of the pioneer work of Heinze (1955), Stegwee and Ponsen (1958), and more recently of Rochow (1963).

The artificial feeding method is carried out by placing aphids that have been starved for several hours in a small feeding chamber. Each feeding chamber is topped with an animal capping-skin membrane previously fastened over the end of a glass tube. Recently, however, the animal membrane has been replaced by stretched Parafilm (Rochow, 1964). The liquid to be tested, partially purified virus suspensions, is placed in the glass tube so that the aphids are able to feed through the membrane top of the feeding chamber into the test solution above it in the tube. Following an acquisition feeding period of about 18 hours at 15°C, the glass tubes are removed from the feeding cage and the aphids are transferred to test plants.

A method for the artificial feeding of individual aphids has been described by Hutchinson and Matthews (1963). The stylets of the aphid are placed in the bore of a small capillary tube, and the aphid is effectively immobilized in this position by securing its legs to the outside of the tube with wax. Using purified solutions of turnip yellow mosaic virus (TYM), heavily labeled with P^{32} and S^{35}, it has been shown that the virus enters the gut of the nonvector aphid *Hyadaphis brassicae* and that substantial quantities of apparently intact virus persist in the gut for periods of days.

Using the needle-injection method, Mueller and Rochow (1961) have regularly transmitted the barley yellow dwarf virus from infected plants to aphids.

The glass needles used in this technique are prepared from 2-mm tubing. Each needle is drawn 3 times over successively smaller flames and a final point is produced by chipping with a razor blade. The completed needles, which have a point diameter of 25–40 μ, are fastened to rubber tubing connected to a rubber syringe bulb. To fill a needle, the operator places its tip in the test solution, presses the syringe bulb with his foot, and allows the solution to rise in the needle. Injection is usually made into the back of the aphid between abdomen and thorax. Liquid is forced into the aphid by pressing with the foot on the syringe bulb. Use of the foot control allows both hands to be used for manipulation of the needle and holding the aphid by means of a brush on the stage of a binocular microscope. If necessary the aphids can be anesthetized with CO_2. Although injected aphids often transmit the virus of barley yellow-dwarf within 1 day, an inoculation test-feeding period of 5 days is usually allowed (Rochow, 1963).

D. Biological Transmission

As used in this context, biological transmission indicates a definite relationship between plant virus and insect vector. We have seen something of this in the discussion of the viruses of potato leaf roll and *Sonchus* yellow vein and their aphid vectors, but it is in regard to some of the leafhopper-borne viruses that our knowledge of this phenomenon is greatest.

All the viruses that are transmitted by leafhoppers and related insects (Cicadellidae) appear to be of the circulative or propagative type; none of the stylet-borne type are known, and, with the exception of potato yellow-dwarf virus, they are not sap-transmissible.

Although the connection of certain plant viruses with leafhoppers has been known in Japan since 1902 and in the United States since 1909, it was not until 1926 that the first evidence was submitted of a biological relationship between the leafhopper and the virus it transmitted. In that year Kunkel, who was a pioneer in this type of work, published an extensive study of aster yellows and pointed out that there was a delay in the development of infective power of at least 10 days, after feeding on an aster plant with "yellows," before the aster leafhopper (*Macrosteles fascifrons* Stal.) became infective to a healthy susceptible plant. Kunkel (1926) interpreted this delay as an "incubation period" while the virus developed in the insect to a sufficient concentration to form an infective dose. At the time this interpretation was not accepted by many workers. Then in 1933 a Japanese worker, Fukushi, demonstrated that the virus of dwarf disease of rice was transmitted to the progeny of the leafhopper vector (*Nephotettix apicalis* var. *cincticeps*), but only if the female insect was viruliferous (virus-bearing). This means, of course, that the virus passed by way of the egg, but not by the sperm, a fact underlined, as we shall see later, by the occurrence of the virus in the cell cytoplasm, and not the nucleus, of the insect vector. Over the years Kunkel carried out a number of experiments designed to show that the virus of aster yellows did in fact multiply in the transmitting leafhopper. In one of these experiments (1937) a number of viruliferous leafhoppers with the aster yellows virus were subjected to a temperature of 32°C for varying periods, and the effect of this on the subsequent ability of the insects to transmit the virus was studied. Kunkel found that the high temperature deprived the leafhoppers of their infectivity for a period, and that the length of this period was dependent on the length of time of their exposure to the high temperature. In other words, leafhoppers kept at 32°C for 1 day regained their infectivity in a few hours; if they were exposed for several days it required 2 days for them to regain infectivity, and exposure for 12 days resulted in a permanent loss of infectivity. Kunkel interpreted these results as indicating that the amount of virus in the insect vector was reduced by the high temperature to a point below infectivity level, and a period of time was therefore necessary for the virus to multiply up again to reach the necessary concentration for infection. Long exposure to heat apparently inactivated all the virus in the insect, which therefore lost infectivity until fed once more on a source of virus. This interpretation was received in 1937, when the results were published, with a good deal of skepticism, but in view of the more direct evidence on plant virus multiplication in leafhoppers it is now generally accepted.

One more experiment of Kunkel's (1955) is of interest in this connection. It is well known that there is a type of acquired immunity among some plant viruses whereby a plant infected with a given virus is resistant to infection by another strain of the same virus, a phenomenon known as cross protection. The current explanation of this is that since all available multiplication sites are occupied by the first virus, there is no place for the second virus to gain access. There are two strains of the aster yellows virus known respectively as the New York and California strains, and the first step was to find a differential host plant which would distinguish between the two strains. Having found such a plant Kunkel demonstrated that infection with one strain precluded the entrance of the other. The insect transmission experiment was carried out as follows: leafhoppers were fed first on a plant infected with New York aster yellows, then transferred to a plant infected with California aster yellows, and finally to the healthy test plant. Next, the experiment was repeated except that this time the order of feeding on the virus sources was reversed. The results were interesting: in every case when the insect was fed successively on the two strains, it transmitted only that on which it fed first. These results clearly show the existence of some biological relationship between insect and virus; it may be that certain cells, possibly of the fat body, were occupied by the strain of virus first imbibed, and this precluded, not necessarily the entrance, but probably the multiplication, of the second virus strain.

These early experiments of Kunkel and others suggested the possibility of plant virus multiplication in insect vectors and laid the foundations for future work.

There are now two main methods of approach to this problem, and a third which is not yet universally applicable. These are the study of transovarial virus transmission, serial transmission of virus from insect to insect by injection, and the visualization of the virus in the insect vector by electron microscopy.

The results of all these methods of study now leave no doubt that some, but not all, leafhopper-transmitted plant viruses do multiply inside the body of the insect vector.

We have seen that in 1933, Fukushi demonstrated that the virus of dwarf disease of rice was transmitted transovarially to the progeny of the vector. Then, in 1940, Fukushi showed that the virus could be passed through six generations involving 82 infective leafhoppers, all derived from a single virus-bearing female without access to a further source of virus. The virus, however, was only passed through the egg and not through the sperm.

Next, Black (1944) discovered the virus of clover club-leaf, not in a plant, but in a leafhopper, *Agalliopsis novella* (Say); he showed (1950) that this virus was transmitted transovarially for more than 5 years through 21 generations from the original viruliferous parents with no further access to a source of virus.

These experiments of both Fukushi and Black prove the multiplication of the virus in the insect, since otherwise the dilution involved would be far too great.

Since these original experiments other examples of leafhopper-borne viruses that are transmitted transovarially have been discussed. For instance, the virus of rice stripe disease has been transmitted in this way through 40 generations of *Delphacodes striatella* (Follen), a planthopper belonging to the Delphacidae (Shinkai, 1955). The same worker (Shinkai, 1960a) demonstrated the transovarial transmission of rice dwarf virus through *Deltocephalus (Inazuma) dorsalis* Motschulski, as well as through *Nephotettix cincticeps,* as originally shown by Fukushi. For further examples of this phenomenon the reader is referred to a recent review by Maramorosch (1963).

The second method of studying the question of virus multiplication in leafhoppers is by serial injection. The pioneers in this approach were Merrill and TenBroeck (1934) who showed that the virus of equine encephalomyelitis multiplied in the mosquito vector, *Aedes aegypti;* they made injection from one mosquito to the next until the dilution was high enough to make multiplication of the virus the only explanation for continued infectivity in the mosquito.

This injection technique was first applied to leafhoppers and plant viruses by Maramorosch (1952), who obtained by this means direct evidence for the multiplication of aster yellows virus in its vector *Macrosteles fascifrons.* Black and Brakke (1952) also made serial transmission of the wound tumor virus in *Agallia constricta* Van Duzee, and proved that this virus, too, multiplied in the leafhopper.

The third method of approach, i.e., the examination of ultrathin sections of viruliferous insects in the electron microscope, has so far only been applied with success in one or two cases. Fukushi *et al.* (1962) and Fukushi and Shikatá (1963) showed the virus of rice dwarf disease *in situ* in the abdomen of the vector *Nephotettix apicalis* var. *cincticeps.* The virus occurred in microcrystals in the cytoplasm of abdominal cells; it was not observed in the cell nuclei or in mitochondria. It is interesting that the microcrystals of virus are somewhat reminiscent of an insect virus, the *Tipula* iridescent virus (TIV) (Smith, 1962), which also occurs in microcrystals in the cell cytoplasm of infected insects; it is an icosahedron as apparently is the virus of rice dwarf disease. Shikata *et al.* (1964) have demonstrated the presence, by means of electron microscopy, of the wound-tumor virus in both the insect vector and the plant tumor (Figs. 45 and 46). Virus particles, obtained from both insect and plant, appeared to be of similar size and morphology. In Fig. 45 can be seen the crystalline arrangement of the virus particles in the fat body of the vector *Agallia constricta.* This arrangement appears similar to that of the virus of rice dwarf disease in the abdomen of the insect vector.

Whitcomb and Black (1961) have made an important contribution in this field by assaying wound-tumor soluble antigen in the insect vector. They measured the rate of increase of wound-tumor soluble antigen at 27.5°C in the vector *Agallia constricta* following injection of massive doses of wound-tumor virus. After injection, soluble antigen was first detected at 4 days, and the soluble antigen titer developed rapidly to a plateau level within 8 to 10 days and was maintained at that level for weeks. A method was developed for determining the presence or absence of soluble antigen in a single insect vector by means of the precipitation ring test. Almost all insects injected with massive doses of wound-tumor virus supported the synthesis of soluble antigen; no injected insects negative for soluble antigen were found to transmit wound-tumor virus to plants.

Sinha and Reddy (1964) have developed an improved fluorescent smear technique for detecting virus antigens in a leafhopper vector. They used wound-tumor virus (WTV)–antiserum conjugates with FITC (crystalline fluorescein isothiocyanate) prepared by a dialysis technique.

In experiments correlating results of smear tests of vectors and their transmission of virus to plants, 107 smears of exposed leafhoppers were stained with the conjugates. Of these, 76 were positive and 24 negative both by smear and transmission tests. Seven leafhoppers were positive by smear test but did not transmit WTV in the transmission tests. On the other hand, there were no leafhoppers which were positive by transmission but negative by smear test.

A quick method of detecting the wound-tumor virus in the leafhopper is by staining the hemolymph smears of the insect with the D(dialysis) conjugates, and observing the color of the fluorescence emitted by them when placed on the oil drop over the condenser of the microscope. The light emitted by smears from viruliferous insects is bright green and that from virus-free insects blue. This method was termed "indirect observation" by Sinha and Black (1963). It is possible, however, to obtain the same effect by observing the colors of smears in ultraviolet light on slides without the use of a microscope. This is called "direct observation" (Sinha and Reddy, 1964).

Since it has now been proved that some plant viruses do multiply in some of their insect vectors, we are faced with the question: Are these agents primarily plant or insect viruses?

From this follows another question: Are the insect vectors which support these viruses themselves affected, either adversely or otherwise?

There is now quite a considerable amount of evidence that this type of insect vector is affected in one way or another. Apart from the presence of the virus itself as observed by means of electron microscopy (Figs. 45, 46) some cytological changes have been recorded. Littau and Maramorosch (1958, 1960) have described changes that occur in the nuclei of fat-body cells of aster leafhoppers carrying the virus of aster yellows. These changes develop in the male leafhoppers 18 to 28 days after acquiring the virus by feeding or injection. In viruliferous insects

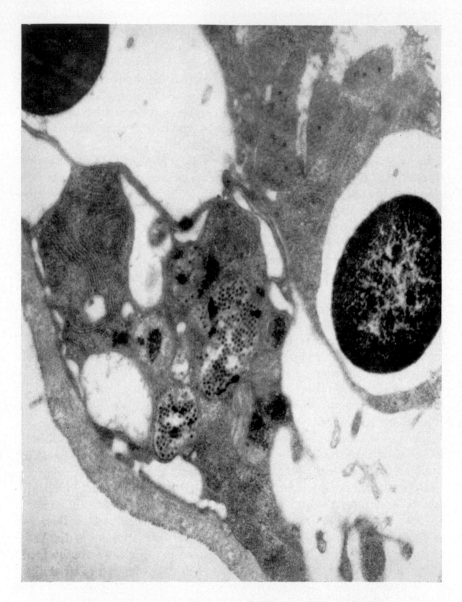

FIG. 45. Clusters of virus particles in fat-body cells from a leafhopper (*Agallia constricta*) infected with the wound-tumor virus. Magnification, × 24,000. (E. Shikata *et al.*, 1964.)

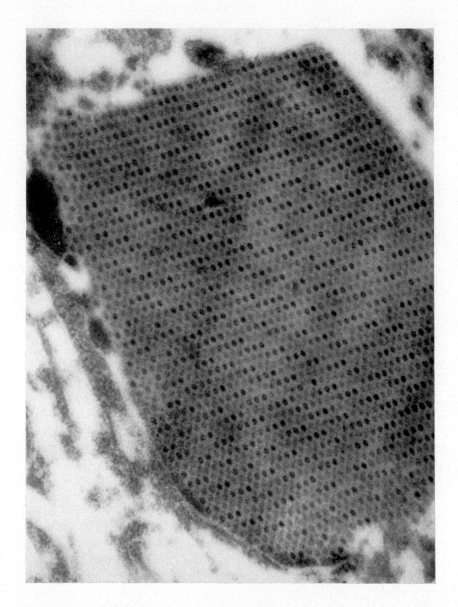

FIG. 46. Crystalline arrangement of wound-tumor virus particles in fat-body cells of *Agallia constricta* (♂). Magnification, × 52,800. (Courtesy E. Shikata.)

the nuclei tend to be stellate rather than rounded and the cytoplasm is reticulate, instead of homogeneous, with many vacuoles. Cytological and histochemical changes in the fat body of *Psammotettix striatus* Fall., the vector of winter wheat mosaic virus, have been described by Lomakina *et al.* (1963; quoted by Maramorosch and Jensen, 1963). In viruliferous insects considerable changes occur in the structure of fat-body cells, similar to those described by Littau and Maramorosch (1960). In addition, a significant reduction was found in the amount of RNA in the cytoplasm and nucleoli, an alteration in the distribution of DNA in the nuclei, and a depletion of polysaccharides in the cytoplasm. These histochemical changes were considered to have been directly responsible for the impaired development and structure of transovarially infected nymphs. Peculiar inclusions were found in the cytoplasm of fat-body cells of infected insects. While no nucleic acids were detected in the inclusions, they did contain histidine.

Changes in the fat-body tissues of the green rice leafhopper, *Nephotettix cincticeps* Uhler, infected with the virus of rice yellow dwarf have been described by Takahashi and Sekiya (1962). About 10 to 15 days following the acquisition of virus, the nuclei of the fat-body cells enlarge and become irregular in shape. After about 20 days, the enlarged nuclei shrink, the irregular shape becomes more pronounced, and vacuoles appear in the cytoplasm. The shrinkage of the nuclei and the cells reaches its maximum after 25 days, the vacuoles increasing in number till the cells appear completely reticulate.

Cytological changes in the green peach aphid, *Myzus persicae* Sulz., infected with the virus of potato leaf roll, have been reported, and less frequently when infected with other viruses (Schmidt, 1959). Chromatin was lacking in some stomach nuclei, and in many instances inclusions were found in the nuclei near the nucleolus.

Sukhov (1940a,b) working with the "Zakuklivanie" or pseudo-rosette disease of oats, described inclusions, or X-bodies, in the cells of the gut, fat-body tissue, and salivary glands of the vector, the leafhopper *Delphacodes (Delphax) striatella* (Fallen). Besides the X-bodies, Sukhov (1940b) also described aciculate crystals of a protein nature in the gut of viruliferous insects and in the cells of infected oat plants. He considered that both the X-bodies and the crystals might be different forms of the pseudo-rosette virus. Similar crystals were observed both in plants affected with winter wheat mosaic and in the intestine of the vector, another leafhopper *Delphocephalus striatus* L. (Sukhov, 1943).

These X-bodies and crystals might be considered analogous to those which occur in tobacco plants affected with tobacco mosaic virus and which have been proved to consist of virus. But since, as we have seen, this virus has no insect vector similar inclusions cannot be observed in an insect.

Needlelike crystals, ranging in size from 3×69 μ to 20×295 μ were found in the intestinal tracts of the leafhopper *Colladonus montanus* (Van Duzee) feeding on celery plants infected with the peach yellow leaf roll strain of Western X disease virus. Crystals were not found in insects propagated on healthy plants, but were present in a high proportion of insects which had fed on diseased plants for 30 or more days. When insects from a diseased stock colony with a high incidence of crystals were transferred to healthy plants, the frequency of individuals with crystals decreased. The crystals were birefringent, and were readily soluble in 0.01 N NaOH as well as in 0.01 M sodium carbonate (pH 10.2) and 0.01 M potassium phosphate (pH 8.0) buffers. They were stable in 0.01 N HCl, and dissolved slowly in distilled water. The origin and nature of the crystals are not known, but the observation that crystalline inclusions do not appear in insects until a sufficient period has been spent feeding on diseased plants may be significant (Lee and Jensen, 1963).

It may be of interest to recall, in this connection, that the citrus red mite *Panonychus citri* (McG.) when infected with a virus disease contains a large number of birefringent crystals. These crystals are symptomatic of the disease and can be used as a diagnostic character (Smith and Cressman, 1962) (see Appendix).

Some alterations in the metabolism of viruliferous insects have been recorded. The virus of dwarf disease of orange causes metabolic changes in the plant hopper vector, *Geisha distinctissima* Wal., resulting in a reduction of both oxygen consumption and total phosphorus (Yoshii and Kiso, 1957a,b). On the other hand, in the rice dwarf disease which is transmitted by the leafhopper *Nephotettix bipunctatus cincticeps* Uhler, oxygen consumption and the respiratory quotient were higher than normal in the viruliferous insects (Yoshii and Kiso, 1959).

Ehrhardt (1960) has studied the effect of the potato leaf roll virus on the metabolism of the aphid vector *Myzus persicae* Sulz. As we have seen earlier, there is good evidence that this virus multiplies in the aphid. Ehrhardt tested the oxygen consumption of nonviruliferous aphids on two types of host plants, *Physalis floridana,* which is susceptible to the leaf roll virus, and Chinese cabbage, which is immune. The oxygen consumption was the same in both cases, 2.95 ml per gm aphids per hour. During an 8-hour feeding period on a source of leaf roll virus only a slight reduction in oxygen consumption occurred. Later, however, during what could be the incubation period of the virus in the insect, oxygen consumption decreased significantly with time, for approximately 30 hours, and later remained constant at approximately 30% reduction. Ehrhardt suggested that the leveling off of oxygen consumption, following 30 hours of gradual reduction, coincides with the completion of the incubation period of the virus in the aphid vector. However, as pointed out by Maramorosch and Jensen (1963), there is no general agreement on the exact duration of the incubation period (Smith, 1931a; Kassanis, 1952).

If, as appears to be the case, some insects are hosts of plant viruses as well as vectors, then some indications of disease or at least of a deleterious effect of the virus on the insect would be expected. Evidence that this is so is now accumulating; the first plant virus shown to be definitely injurious to the insect vector was that causing Western X disease of stone fruits and a yellows disease in celery. This virus has been shown to cause a significant reduction in the life span of the vector *Colladonus montanus* (Van Duzee). The mean longevity for 116 transmitting individuals of *C. montanus* was 20 days, whereas 64 nontransmitting insects in the same experiment averaged 51 days. The criticism might be made that the shortened life period was due to a difference in the nutritional value or digestibility of the virus-infected plant rather than to the virus itself. However, that variable has been eliminated because it was shown that the acquisition of the virus by the insect was the essential factor. In other words, if, for some reason, insects failed to acquire virus, such nontransmitting insects lived as long on the virus-infected plants as did the control insects on the virus-free plants.

There is also a reduction in the number of eggs laid by viruliferous females of *C. montanus;* in a carefully controlled experiment it was found that infective stocks laid only one-third the number of eggs laid by the healthy controls in equal time periods (Jensen, 1958, 1959a,b, 1962, 1963).

Evidence is accumulating of other similar cases; Shinkai (1960b) reports the apparently deleterious effect of rice dwarf virus on nymphs of *Inazuma dorsalis.*

The mean adult longevity of viruliferous individuals of *N. cincticeps,* a vector of rice dwarf disease, was 19–20 days compared to 29 days for the healthy controls (Satomi *et al.*, personal communication in Maramorosch and Jensen, 1963).

A somewhat surprising effect of a plant virus upon a nonvector insect has been described by Maramorosch (1957, 1958a,b). The leafhopper, *Dalbulus maidis* (DeLong and Walcutt), can acquire and retain aster yellows virus in its hemolymph, but it cannot transmit the virus. The insect is highly host-specific and survives and breeds well only on maize (*Zea mays*) and on the related wild grass teosinte (*Euchlaena mexicana*). It dies within 4 days on most other plants, including healthy China asters (*Callistephus chinensis*). However, if the aster plants are infected with aster yellows, the insects may survive for long periods, either as nymphs or adults. What is more surprising, however, is the fact that after the insects have survived for 7 days and longer on diseased aster plants, they are then able to survive on healthy asters and on other plants such as carrots and rye (Maramorosch, 1960).

Although a number of experiments have been carried out in efforts to explain this phenomenon, the mechanism responsible for the increased survival of virus-carrying *D. maidis* is not yet known. For a fuller account of the harmful and beneficial effects of plant viruses on insects the reader should consult a review by Maramorosch and Jensen (1963).

Before concluding the discussion on biological transmission of plant viruses mention must be made of a virus disease of sugarbeet known as "beet savoy" in the United States, and of a very similar disease known as "beet leaf curl" in Germany. The vector in both cases is somewhat unusual, it is not a leafhopper but a "lace bug," so called, belonging to the *Piesmidae; Piesma cinerea* in the United States, and *P. quadrata* in Germany. The American disease has been studied by Coons *et al.* (1958) and by Von Volk and Krczal (1957) in Germany. There is a good deal of evidence of a biological relationship between virus and vector, particularly in the length of time the virus is retained by the insect, apparently right through the hibernation period of the adult.

A recent paper Proeseler (1964) affords further evidence on this biological relationship, the author carried out 4 serial passages of the virus from insect to insect by means of carefully prepared micropipettes, but lost all adult insects in the fifth passage. From a personal communication (1964) from the author to Karl Maramorosch, it appears that a mysterious mortality several times prevented continuous serial passages and often the second or third group suddenly died. Maramorosch suggests

Black, L. M., and Brakke, M. K. (1952). *Phytopathology* **42**, 269–273.

Bradley, R. H. E. (1952). *Ann. Appl. Biol.* **39**, 78.

Bradley, R. H. E. (1956). *Can. J. Microbiol.* **2**, 539.

Bradley, R. H. E. (1959). *Proc. 7th, Intern. Botan. Congr. Montreal, 1960.*

Brčák, J. (1957). *Phytopathol. Z.* **30**, 414–428.

Brčák, J. (1959). *Virusol.* **8**, 171–176.

Chaudhuri, R. P. (1950). *Ann. Appl. Biol.* **37**, 342.

Clinch, P., Loughnane, J. B., and Murphy, P. (1936). *Sci. Proc. Roy. Dublin Soc.* **21**, 431.

Coons, G. H., Dewey, S., Bockstahler, H. W., and Schneider, C. L. (1958). *Plant Disease Reptr.* **42**, 502–511.

Cornuet, P., and Morand, J. D. (1960a). *Compt. Rend.* **250**, 1583–1584.

Cornuet, P., and Morand, J. D. (1960b). *Compt. Rend.* **250**, 1750–1752.

Costa, A. S., de Silva, D. M., and Duffus, J. E. (1958). *Virology* **5**, 145–149.

Day, M. F. (1955). *Exptl. Parasitol.* **4**, 387.

Day, M. F., and Irzykiewicz, H. (1954). *Australian J. Biol. Sci.* **7**, 251.

Duffus, J. E. (1960). *Phytopathology* **50**, 289.

Duffus, J. E. (1963). *Virology* **21**, 194–202.

Ehrhardt, P. (1960). *Entomol. Exptl. Appl.* **3**, 114–117.

Elmer, O. H. (1925). *Iowa Res. Bull.* **82**, 91.

Fukushi, T. (1933). *Proc. Imp. Acad. (Tokyo)* **9**, 451.

Fukushi, T. (1940). *J. Fac. Agr., Hokkaido Univ.* **45**, 83–154.

Fukushi, T., and Shikata, E. (1963). *Virology* **21**, 503.

Fukushi, T., Shikata, E., and Kimura, I. (1962). *Virology* **18**, 192–205.

Gamez, R., and Watson, M. A. (1964). *Virology* **22**, 292–295.

Heinze, K. (1955). *Phytopathol. Z.* **25**, 103.

Hoggan, I. A. (1931). *Phytopathology* **21**, 199–212.

Hoggan, I. A. (1934). *J. Agr. Res.* **49**, 1135–1142.

Hutchinson, P. B., and Matthews, R. E. F. (1963). *Virology* **20**, 169–175.

Jensen, D. D. (1958). *Phytopathology* **48**, 394 (abstr.).

Jensen, D. D. (1959a). *Virology* **8**, 164–175.

Jensen, D. D. (1959b). *Pan-Pacific Entomologist* **35**, 65–82.

Jensen, D. D. (1962). *Proc. 11th, Intern. Congr. Entomol., Vienna, 1960* pp. 790–791.

Jensen, D. D. (1963). *Ann. N.Y. Acad. Sci.* **105**, 685–712.

Jensen, D. D., Griggs, W. H., Gonzales, C. G., and Schneider, H. (1964). *Phytopathology* **54**, 1346, 1351.

Kassanis, B. (1947). *Ann. Appl. Biol.* **34**, 312.

Kassanis, B. (1952). *Ann. Appl. Biol.* **39**, 157.

Kassanis, B. (1961). *Virology* **13**, 93.

Kassanis, B. (1963). *Advan. Virus Res.* **10**, 219–255.

Kennedy, J. S., Day, M. F., and Eastop, V. F. (1962). "*A Conspectus of Aphids as Vectors of Plant Viruses,*" 114 pp. Commonwealth Inst. Entomol., London.

Kikumoto, T., and Matsui, C. (1962). *Virology* **16**, 509–510.

Kunkel, L. O. (1926). *Am. J. Botany* **13**, 646.

Kunkel, L. O. (1937). *Am. J. Botany* **24**, 316.

Kunkel, L. O. (1955). *Advan. Virus Res.* **3**, 251.

Kvíčala, B. (1945). *Nature* **155**, 174.

Lee, P. E., and Jensen, D. D. (1963). *Virology* **20**, 328–332.

Littau, V. C., and Maramorosch, K. (1958). *Phytopathology* **48**, 263.

Littau, V. C., and Maramorosch, K. (1960). *Virology* **10**, 483–500.

Lomakina, L. Ya., Razvyazkina, G. M., and Shubnikova, E. A. (1963). *Vopr. Virusol.* *XI*, 168–172.

Maramorosch, K. (1952). *Phytopathology* **42**, 59–64.

Maramorosch, K. (1957). *Phytopathology* **47**, 23.

Maramorosch, K. (1958a). *Proc. 7th Intern. Congr. Microbiol., Stockholm, 1958* p. 260. Almqvist & Wiksell, Uppsala.

Maramorosch, K. (1958b). *Tijdschr. Plantenziekten* **64**, 383–391.

Maramorosch, K. (1960). *Sci. Am.* **203**, 138–144.

Maramorosch, K. (1963). *Ann. Rev. Entomol.* **8**, 369–414.

Maramorosch, K. (1964). Personal communication.

Maramorosch, K., and Jensen, D. D. (1963). *Ann. Rev. Microbiol.* **17**, 495–530.

Merrill, H. M., and TenBroeck, C. (1934). *Proc. Soc. Exptl. Biol. Med.* **32**, 421.

Mueller, W. C., and Rochow, W. F. (1961). *Virology* **14**, 253–258.

Nault, L. R., Gyrisco, G. C., and Rochow, W. F. (1964). *Phytopathology* **54**, 1269–1272.

Newton, W. (1953). *FAO Plant Protect. Bull.* **2**, 40.

Nishi, H. (1962). *Program Abstr. Symp. Plant Viruses, Tokyo, 1962.*

Olitsky, P. K. (1925). *Science* **62**, 442.

Orlob, G. B. (1962). *Virology* **16**, 301–304.

Orlob, G. B. (1963). *Phytopathology* **53**, 822–830.

Orlob, G. B., and Arny, D. C. (1960). *Virology* **10**, 273–274.

Osborn, H. T. (1935). *Phytopathology* **25**, 160.

Ossiannilsson, F. (1958). *Kgl. Lantbruks-Hogskol. Ann.* **24**, 369–374.

Oswald, J. W., and Houston, B. R. (1953). *Phytopathology* **43**, 128–136.

Paine, J., and Legg, J. T. (1953). *Nature* **171**, 263–264.

Prentice, I. W., and Woollcombe, T. M. (1951). *Ann. Appl. Biol.* **33**, 50.

Proeseler, G. (1964). *Z. Angew. Entomol.* **54**, 325–333.

Rochow, W. F. (1959). *Plant Disease Reptr.* Suppl. 262, 356–359.

Rochow, W. F. (1960a). *Phytopathology* **50**, 881–884.

Rochow, W. F. (1960b). *Plant Disease Reptr.* **44**, 940–942.

Rochow, W. F. (1961). *Phytopathology* **51**, 809–810.

Rochow, W. F. (1963). *Ann. N.Y. Acad. Sci.* **105**, 713–729.

Rochow, W. F. (1964). Personal communication.

Rochow, W. F., and Pang, E.-Wa. (1961). *Virology* **15**, 382–384.

Schmidt, H. B. (1959). *Biol. Zentr.* **78**, 889–936.

Schmidt, H. B. (1960). *Monatsber. Deut. Akad. Wiss. Berlin* **2**, 214–223.

Schmutterer, H. (1961). *Z. Angew. Entomol.* **47**, 277–301 and 416–439.

Shikata, E., Orenski, S. W., Hirumi, H., Mitsuhashi, J., and Maramorosch, K. (1964). *Virology* **23**, 441–444.

Shinkai, A. (1955). *Ann. Rept. Kanto Tosan Phytopathol. Entomol. Soc.* pp. 5–6 (in Japanese).

Shinkai, A. (1960a). *Plant Protect.* **14**, 146–150 (in Japanese).

Shinkai, A. (1960b). *Ann. Phytopathol. Soc. Japan* **25**, 42.

Simons, J. N. (1954). *Phytopathology* **44**, 282.

Sinha, R. C., and Black, L. M. (1963). *Virology* **21**, 181–187.

Sinha, R. C., and Reddy, D. V. R. (1964). *Virology* **24**, 626–634.

Slykhuis, J. T. (1953). *Can. J. Agr. Sci.* **33**, 195–197.

Slykhuis, J. T. (1955). *Phytopathology* **45**, 116–128.

Slykhuis, J. T., Zillinsky, R. J., Hannah, A. E., and Richards, W. R. (1959). *Plant Disease Reptr.* **43**, 849–854.

Smith, K. M. (1931a). *Ann. Appl. Biol.* **18**, 141.

Smith, K. M. (1931b). *Proc. Roy. Soc.* **B109**, 251–267.

Smith, K. M. (1941). *Parasitology* **33**, 110–116.

Smith, K. M. (1946). *Parasitology* **37**, 126.

Smith, K. M. (1962). *Advan. Virus Res.* **9**, 195–240.

Smith, K. M., and Cressman, A. W. (1962). *J. Insect Pathol.* **4**, 229–236.

Stegwee, D., and Peters, D. (1961). *Virology* **15**, 202–203.

Stegwee, D., and Ponsen, M. B. (1958). *Entomol. Exptl. Appl.* **1**, 291.

Stubbs, L. L. (1955). *Australian J. Biol. Sci.* **8**, 68–74.

Sukhov, K. S. (1940a). *Mikrobiologiya* **9**, 188–196.

Sukhov, K. S. (1940b). *Compt. Rend. Acad. Sci. URSS* **27**, 377–379 (in Russian).

Sukhov, K. S. (1943). *Compt. Rend. Acad. Sci. URSS* **39**, 73–75.

Sukhov, K. S. (1944). *Compt. Rend. Acad. Sci. URSS* **42**, 226–228.

Sylvester, E. S. (1949). *Phytopathology* **39**, 117.

Sylvester, E. S. (1954). *Hilgardia* **23**, 53.

Sylvester, E. S. (1962). Mechanisms of plant virus transmission by aphids. *In* "Biological Transmission of Disease Agents" (K. Maramorosch, ed.), pp. 11–31. Academic Press, New York.

Takahashi, Y., and Sekiya, I. (1962). *Ann. Appl. Zool. Entomol.* **6**, 90–94.

Toko, H. V., and Bruehl, G. W. (1956). *Plant Disease Reptr.* **40**, 284–288.

Toko, H. V., and Bruehl, G. W. (1957). *Phytopathology* **47**, 536 (abstr.).

Toko, H. V., and Bruehl, G. W. (1959). *Phytopathology* **49**, 343–347.

van der Want, J. P. H. (1954). "Onderzoekingen over Virusziekten van der Boon (*Phaseolus vulgaris* L.)," 6 pp. Veenman, Wageningen.

van Hoof, H. A. (1957). *Koninkl. Ned. Akad. Wetenschap., Proc.* **C60**, 314.

van Soest, W., and de Meesters-Manger Cats, V. (1956). *Virology* **2**, 411–414.

Volk, Von J., and Krczal, H. (1957). *Nachrbl. Deut. Pflanzenschutzdienst (Berlin)* [N.S.] **9**, 17–22.

Walters, H. J. (1952). *Phytopathology* **42**, 355.

Watson, M. A. (1936). *Phil. Trans. Roy. Soc. London* **B226**, 457.

Watson, M. A. (1938). *Proc. Roy. Soc.* **B125**, 144.

Watson, M. A. (1946). *Proc. Roy. Soc.* **B133**, 200.

Watson, M. A. (1962). *Ann. Appl. Biol.* **50**, 451.

Watson, M. A., and Mulligan, T. (1960). *Ann. Appl. Biol.* **48**, 711–720.

Watson, M. A., and Roberts, F. M. (1939). *Proc. Roy. Soc.* **B127**, 543.

Whitcomb, R. F., and Black, L. M. (1961). *Virology* **15**, 136–145.

Yoshii, H., and Kiso, A. (1957a). *Virus* **7**, 306–314.

Yoshii, H., and Kiso, A. (1957b). *Virus* **7**, 315–320.

Yoshii, H., and Kiso, A. (1959). *Virus* **9**, 415–422.

CHAPTER XII

Viruses and the Biological Control
of Insect Pests

Introduction

Biological control, as it is usually called, has been used widely, not always with success. The earliest efforts in this direction employed mainly predacious or parasitic insects, which were produced in large numbers under laboratory conditions and then introduced into the area where the pest in question was active.

Biological control, however, is not confined to the encouragement of predatory insects, and more attention is now being paid to the possibility of starting epizootics among agricultural pests, or, in other words, microbiological control. Both bacterial and fungal disease agents have been tried and some success has been obtained with a bacterial disease of the Japanese beetle, *Popillia japonica* Newn., by producing the spores of *Bacillus popilliae* Dutky, the causative agent of the so-called "milky disease," in large quantities and disseminating them in the soil against the beetle larvae in the eastern United States. One of the crystal-forming bacteria, *Bacillus thuringiensis* Berliner, has been, and is being, used extensively in the control of harmful insects. The crystals formed by the bacilli are very toxic to a large number of insects, so that in a sense this is more a case of an insecticidal, rather than a microbial, control.

The idea of using viruses in biological control is not new, the classical example being the introduction of the myxoma virus into Australia against the rabbit pest, and, more recently, into France and Great Britain. What is more recent is the use of viruses to control insect pests; that this has not been developed earlier is not surprising in view of the long neglect of the study of insect viruses. However, this omission is now being remedied, and, as we shall see later, quite a number of experiments using viruses as control agents have been carried out.

A. Selection of Viruses

This, of course, obviously depends on what virus, or viruses, are available for use against a particular insect pest. However, where a choice exists, it is better to use one of the inclusion-body viruses and the more virulent the better. The main advantage in using this type of virus lies in the inclusion body itself, whether polyhedron or capsule (granule). Here, the virus is already packaged for use, so to speak; it is protected against adverse environmental conditions, and, in the case of the polyhedroses, is in a convenient form for virus assay. Furthermore, the inclusion body prolongs the infectivity of the virus for long periods, even for years, and thus ensures that it will persist in the vicinity of the insect pest.

The effects of the virus on its host are also important; in other words the kind of disease developed may affect the ease of spread. For this reason the nuclear polyhedroses are likely to be more effective for control measures than the cytoplasmic polyhedroses. In the former, the destruction of the skin and the liquefaction of the body contents, containing millions of polyhedra, ensure the liberation and distribution of the polyhedra. In the cytoplasmic polyhedroses the skin is not destroyed, nor are the body contents liquefied. The polyhedra are therefore less accessible though some are liberated by regurgitation and by passage with the feces.

The granuloses are usually as effective in control measures as are the nuclear polyhedroses, since the two diseases are very similar, with disruption of the skin and liberation of the liquefied body contents, filled either with polyhedra or capsules (granules) as the case may be (Fig. 19).

B. Variable Factors

It is clear that one important factor in the use of viruses as insecticides is the accessibility of the insect pest to be controlled. Larvae which feed gregariously and openly on the foliage of the host plant are most easily approached. The most promising results have been obtained against insects of this type, notably two species of sawfly larvae, the cabbage looper, and the two species of cabbage worm. On the other hand, insects living in a concealed habitat are more difficult to cope with, such are the various budworms and soil-living insects.

An important factor to be considered is the length of the incubation period of the virus in the insect. If this is too prolonged the purpose of the virus application will be defeated. Even if there is a 100% kill of the larvae, the result will not be satisfactory if there is sufficient time intervening between application of virus and death of the insect to allow feeding damage. For example, a virus spray failed to prevent fruit damage on apple trees by the red-banded leaf-roller, *Argyrotaenia velutinana* (Walker), in spite of nearly 100% mortality (Glass, 1958). Similarly, application of a polyhedrosis virus to the tobacco budworm gave 100% mortality in 13 days, but this was not satisfactory because much feeding damage took place during the 10 days required for the virus to take effect (Chamberlin and Dutky, 1958).

There are certain factors which may govern to some extent the duration of the incubation period. Temperature and humidity are important because, although they do not affect susceptibility, they do affect the time the virus takes to act. Low temperatures prolong the incubation period and may also affect the population densities.

Rain does not appear to have much effect on the efficiency of a virus application, once the polyhedra have dried on the foliage they are not easily removed, especially if a sticking agent is used (Thompson and Steinhaus, 1950; Bird, 1964; Burgerjon and Grison, 1965).

The concentration of the virus, or, in other words, the number of polyhedra, the number of applications, and the time of application are important factors to take into consideration.

The optimum concentration of a virus varies with the particular virus and must usually be arrived at by trial and error.

In the control of the cabbage looper *Trichoplusia ni* (Hübner) it was found, in laboratory tests, that an oral dosage of 0.001 ml of an inoculum prepared by triturating the body contents of one virus-killed fifth-instar larva in 16 liters of water caused infection and death of all larvae treated.

Four field tests on brassica crops were carried out, using suspensions varying from 0.94 to 120 larvae per acre applied with a low-volume sprayer at the rate of 30 gallons per acre. In all four experiments a polyhedrosis epizootic was initiated, greatly reducing the looper population; but the increased dosage caused more rapid disease progress. Hemocytometer counts indicated that the lowest spray dilution contained 0.6 million and the highest 7.6 million polyhedra per acre (McEwen and Hervey, 1958).

In some experiments with a nuclear polyhedrosis against the same cabbage looper as a pest of lettuce crops, it was found that treatments with low virus concentrations prolonged the incubation period as did also low temperatures. Two applications of a single concentration were more effective than one, and were as effective as a single application of a higher concentration (Hall, 1957).

The time of application of the virus is important and this is intimately associated with larval susceptibility. As a rule the early larval instars are more susceptible to infection; also the smaller the larvae the smaller will be the leaf damage. Experiments carried out by Elmore (1961) on the cabbage looper illustrate this. A virus concentrate of 5.5 billion polyhedra per milliliter was used at the rates of 1, 5, and 10 ml per gallon. In April, the time required from exposure of hatching larvae until most of them died was 20 days. In September, cabbage loopers were not successfully controlled, nor was leaf damage prevented, after two applications were made to half-grown cauliflowers when loopers had become large enough to cause some leaf damage before dying. However, virus applications on small plants soon after thinning, when loopers were small or in the egg stage, gave much better control. Three applications protected a cauliflower field from plant thinning until harvest time.

It is probably true to say that doses of high virus concentration, started early and repeated frequently, are the most likely to be successful. In a study of the effectiveness of the nuclear polyhedrosis for the control of the cabbage looper, Getzin (1962) arrived at the following conclusions. High doses of polyhedra (9.5×10^{11} per acre) were more effective than low doses (9.5×10^9 per acre), although all rates caused significant larval mortality. Weekly applications were more effective than biweekly applications. The latter treatments did not give significant control until 39 days after the initiation of the spray.

C. Application of the Virus

The virus preparations can be applied either as a dust or as a spray, and most of the standard spraying or dusting machines used for the application of insecticides are suitable. Probably a liquid spray gives better coverage. For liquid preparations a low-volume sprayer is best.

Wetting agents or stickers seem to have no deleterious effect on the viruses and may be used to increase the wetting and adhesive power of the virus material. For example, suspensions of cytoplasmic polyhedra from the armyworm *Thaumetopoea pityocampa* Schiff. when mixed with 0.2% of a wetting agent caused high mortality and persisted on pine foliage under laboratory conditions when exposed to simulated rainfall (Burgerjon and Grison, 1965).

It has been found that a combination of a virus and another pathogen or insecticide is sometimes more effective than the virus acting alone. Steinhaus (1951) found a combination of nuclear polyhedrosis virus and *Bacillus thuringiensis* Berliner more efficient than the virus alone to control the larvae of the alfalfa butterfly.

Wolfenbarger (1965) has studied polyhedrosis-virus-surfactant and insecticide combinations, for the control of the cabbage looper. He found that the use of an oil and endrin, when added to a virus-water solution, increased the control of infestations on the waxy surface of the cabbage leaf. A threefold increase in virus dosage was significantly inferior to the virus, oil, and endrin combinations.

Similarly in experiments to control the Great Basin tent caterpillar, *Malacosoma fragile* (Stretch), a combination of a nuclear polyhedrosis virus and *Bacillus thuringiensis* Berliner provided a more efficient means of control than either agent acting alone (Stelzer, 1965).

When spraying or dusting the foliage, complete coverage of both upper and lower leaf surfaces is essential. When first hatched, young larvae frequently feed on the lower surface, eating only the epidermis. Thus, they may not come into contact with the virus on the upper surface for a period of several days or even longer.

D. Preparing and Storing the Virus

In preparing virus in quantity for use in microbial control, one fundamental characteristic of viruses must be reckoned with. Multiplication can only take place inside a living susceptible cell. Now as we have already learned, neither the insect nor its food plant is absolutely necessary. In other words it is, or soon will be, possible to propagate the virus in tissue culture without the necessity of raising and infecting large numbers of larvae. This technique is not yet generally available. It is, however, possible to dispense with the plant in some cases and raise the necessary larvae on a synthetic or semi-synthetic diet (see Chapter X,E). This has already been achieved in studies on the effectiveness of a nuclear polyhedrosis virus against field populations of the bollworm *Heliothis zea* (Boddie) and the tobacco budworm *H. virescens* Fabr.

In two tests the virus compared favorably with the insecticides recommended for use on cotton (Ignoffo et al., 1965).

In the absence of synthetic media there is no alternative to raising large numbers of larvae of the pest in question and infecting them artificially with the virus. Sometimes it is possible to take advantage of a natural epizootic to collect large quantities of infected insects and thus lay the foundation of a virus stock. This was done by Steinhaus and Thompson (1949) in field tests with a nuclear polyhedrosis virus against the alfalfa caterpillar. The caterpillars were gathered from naturally infected fields, brought into the laboratory and placed within large barriers or rearing trays where they were crowded together with previously infected individuals. In a few days nearly all were dead or dying from the polyhedrosis. Their infected bodies were collected periodically, placed in large glass vials, and then held in the refrigerator until needed.

Preliminary clarification and centrifugation of the polyhedra and/or capsules enable them to be stored at 4°C for long periods while the build-up of the virus stocks proceeds. This, of course, is one advantage of the use of the inclusion-body viruses in microbial control.

The bodies of the infected caterpillars are mixed with distilled water and well triturated, preferably with a Waring blendor. The material can then be further diluted with distilled water and filtered through cheesecloth to remove the larger particles. One or two cycles of centrifugation will give a fairly pure suspension of the inclusion bodies.

E. Standardization of Virus Preparations

In small-scale efforts to control a local insect pest by means of a virus, it is not necessary to have a highly purified product or one which is carefully standardized. A crude example is the use of 5 fully grown larvae of *Pieris brassicae*, infected with granulosis, to a gallon of water to make an infectious spray.

For more accurate standardization, counts can be made by means of a hemocytometer of the actual number of polyhedra or capsules, and a rough calculation made from this of the total number of polyhedra per milliliter. However, even so this is still only an approximation since such a count does not take into consideration the virus particles in the suspension not occluded in polyhedra or capsules.

If a noninclusion virus, such as TIV, is to be used in control, some idea of concentration can be arrived at by actual particle-counts in the electron microscope or by the degree of light scattering in the virus suspension.

A method of bioassay of virus activity of the nuclear polyhedrosis of *Heliothis* using a semi-synthetic diet has been described by Ignoffo (1965). The hot liquid diet was poured into 8-dram glass shell vials to a depth of about 40 mm. After solidification, the diet was contaminated on the surface with polyhedral inclusion bodies (PIBs) dispensed in 0–1-ml volumes by means of 1.0-ml serological pipettes calibrated in 0.01-ml units. The PIBs dilutions were made from a water suspension of the virus averaging $1.111 \pm 9 \times 10^5$ PIBs/ml (six samples) and then later transformed to number of PIBs/mm^2 of diet surface. A ball-point glass rod was used to distribute the virus uniformly over the entire surface of the diet. One larva was placed in each vial after the surface of the diet dried. The vials were then plugged with cotton batting, placed at 29°C, and examined daily for larval mortality. All tests were terminated after 10 days since by that time the larvae had either died or pupated. The activity measurement of PIBs may be terminated within 5 days, instead of the usual ten, if lethal time is used instead of lethal dose. After 5 days' exposure the larval mortality for 3, 15, 29, 58, 146, and 209–29,200 PIBs/mm^2 was 4, 22, 34, 50, 67, and 82%, respectively.

The results of the experiments were as follows. The LD$_{50}$ and 95% fiducial limits were 32.0 and 26.2–39.0 PIBs/mm^2 of diet, respectively. The LT$_{50}$ values for 29,200, 292, 146, 58, and 29 PIBs/mm^2 were 3.8, 3.8, 4.3, 5.3, and 6.5 days, respectively.

In a recent work conference held to discuss and review the progress of research on, registration, and standardization of, the nucleopolyhedrosis virus of *Heliothis zea* (Boddie) a number of resolutions were made (Falcon, 1965). It was first pointed out the virus provides a potential apparently safe substitute for several hazardous pesticides. Among other resolutions it was emphasized that the virus materials destined for commercial use be standardized in content and insecticidal activity, in order that a highly effective, safe, and economical insecticidal product may be obtained.

In starting what may eventually be a new industry it is obviously important that, at the beginning, there should not be a number of virus preparations, whether for *Heliothis* or other pests, which have been hurriedly prepared without adequate testing or standardization.

F. Some Examples of Virus Control

There are several examples of the successful use of viruses in the control of insect pests, but with the exception of *Heliothis* spp., mentioned above, they are mainly insects which feed openly on the food plant.

As we have already seen, a lot of experiments have been carried out on the control of the cabbage looper, mainly on cauliflower and lettuce, by means of a nucleopolyhedrosis, and these have been fairly successful (Hall, 1957; McEwen and Hervey, 1958; Elmore, 1961).

Some success has been achieved in Egypt with a nuclear polyhedrosis virus against the cotton leafworm *Prodenia litura* Fabr. on cotton and other field crops (Abul-Nasr, 1959).

Granulosis viruses have been used against the larvae of *Pieris brassicae* (Biliotti *et al.*, 1956), and against the larvae of *Pieris rapae* (Kelsey, 1957) both on brassica crops.

Perhaps the most outstanding successes have been made with a nuclear polyhedrosis against the larvae of pine sawflies *Neodiprion sertifer*, and of the spruce sawfly, *Diprion hercyniae*. Successful control of the wattle bagworm in South Africa has been achieved with a nuclear polyhedrosis virus. The insect is indigenous to South Africa, where it lives on a species of *Acacia*, from which it has invaded plantations of black wattle (*A. mollissima* Willd.)—a crop of considerable importance. Aqueous suspensions of 10,000 polyhedra/mm^3 caused a very high mortality. Such suspensions remain effective for a considerable time, so that they can be applied some months before the bagworms hatch (Ossowski, 1959).

As regards the effectiveness of viruses against *Heliothis* spp., Ignoffo *et al.* (1965) report the successful use of a nuclear polyhedrosis virus against field populations of the bollworm *H. zea* and the tobacco budworm *H. virescens* Fabr. on cotton.

This virus is the subject of the report of the work conference for the registration and standardization of virus insecticides previously mentioned (Falcon, 1965).

For a discussion on the general principles of biological control the reader is referred to DeBach (1964).

REFERENCES

Abul-Nasr, S. (1959). Further tests on the use of a polyhedrosis virus in the control of the cotton leafworm *Prodenia litura* Fabr. *J. Insect Pathol.* **1**, 112–120.

Biliotti, E., Grison, P., and Martouret, D. (1956). L'utilization d'une maladie à virus comme méthode de lutte biologique contre *Pieris brassicae* L. *Entomophaga* 1, 35–44.

Bird, F. T. (1964). The use of viruses in biological control. *Intern. Colloq. Insect Pathol. Lutte Microbiol.*, 2ᵉ, *Paris, 1962* p. 465–473.

Burgerjon, A., and Grison, P. (1965). Adhesiveness of preparations of *Smithiavirus pityocampae* Vago on pine foliage. *J. Invert. Pathol.* 7, 281–284.

Chamberlin, F. S., and Dutky, S. R. (1958). Tests of pathogens for the control of tobacco insects. *J. Econ. Entomol.* 51, 560.

DeBach, P., ed. (1964). "Biological Control of Insect Pests and Weeds." Reinhold, New York.

Elmore, J. C. (1961). Control of the cabbage looper with a nuclear polyhedrosis disease. *J. Econ. Entomol.* 54, 47–50.

Falcon, L. A. (1965). Report of work conference on the utilization of nucleopolyhedrosis virus for the control of *Heliothis zea*. *Bull. Entomol. Soc. Am.* 11, 84.

Getzin, L. W. (1962). The effectiveness of the polyhedrosis virus for control of the cabbage looper, *Trichoplusia ni*. *J. Econ. Entomol.* 55, 442–445.

Glass, E. H. (1958). Laboratory and field tests with the granulosis of the red banded leaf roller. *J. Econ. Entomol.* 51, 454–457.

Hall, I. M. (1957). Use of a polyhedrosis virus to control the cabbage looper on lettuce in California. *J. Econ. Entomol.* 50, 551–553.

Ignoffo, C. M. (1965). The nuclear polyhedrosis virus of *Heliothis zea* (Boddie) and *Heliothis virescens* (Fabr.). IV. Bioassay of virus activity. *J. Invert. Pathol.* 7, 315–319.

Ignoffo, C. M., Chapman, A. J., and Martin, D. F. (1965). The nuclear polyhedrosis virus of *Heliothis zea* (Boddie) and *Heliothis virescens* (Fabr.). III. Effectiveness of the virus against field populations of *Heliothis* on cotton, corn and grain sorghum. *J. Invert. Pathol.* 7, 227–235.

Kelsey, J. M. (1957). Virus sprays for control of *Pieris rapae* L. *New Zealand J. Sci Technol.* A38, 644–646.

McEwen, F. L., and Hervey, G. E. R. (1958). Control of the cabbage looper with a virus disease. *J. Econ. Entomol.* 51, 626–631.

Ossowski, L. L. J. (1959). The use of a nuclear virus disease for the control of the wattle bagworm, *Kotochalia junodi* (Heyl.). *Proc. 4th Intern. Congr. Crop Protect., Hamburg, 1957* Vol. 1, pp. 879 and 883.

Steinhaus, E. A. (1951). Possible use of *Bacillus thuringiensis* Berliner as an aid in the biological control of the alfalfa caterpillar. *Hilgardia* 20, 359–381.

Steinhaus, E. A., and Thompson, C. G. (1949). Preliminary field tests using a polyhedrosis virus to control the alfalfa caterpillar. *J. Econ. Entomol.* 42, 301–305.

Stelzer, M. J. (1965). Susceptibility of the Great Basin tent caterpillar *Malacosoma fragile* (Stretch) to a nuclear polyhedrosis virus and *Bacillus thuringiensis* Berliner. *J. Invert. Pathol.* 7, 122–125.

Thompson, C. G., and Steinhaus, E. A. (1950). Further tests using a polyhedrosis virus to control the alfalfa caterpillar. *Hilgardia* 19, 411–445.

Wolfenbarger, D. A. (1965). Polyhedrosis-virus-surfactant and insecticide combinations, and *Bacillus thuringiensis* surfactant combinations for cabbage looper control. *J. Invert. Pathol.* 7, 33–38.

Appendix

Arthropoda: Arachnida

Two viruses, and a possible third, have been described in members of the Arthropoda beside the order Insecta. These are the spider mites, Arachnida; and a brief account is given of their viruses and the diseases caused.

Panonychus citri (McGregor). The Citrus Red Mite

The disease in this mite is very infectious and spreads readily in colonies of mites. The symptoms consist of paralysis with the legs stiffened ventrally in a stiltlike manner. Diarrhea frequently occurs and the mites may be found dead with the anal end fixed to the feeding surface by fecal material.

The virus is best isolated by differential centrifugation, the final spins being made at 36,000 rpm (11,000 g) on a "Spinco" centrifuge. The virus is very small, measuring about 35 mμ in diameter; its six-sided contour suggests that it is an icosahedron (Smith *et al.*, 1959).

A curious feature of this disease is the almost invariable appearance, in the body and legs of the infected mite, of large numbers of birefringent crystals. These are not virus crystals but appear to be the result of a disordered metabolism.

In sections of diseased mites, either stained or unstained, numerous dark round bodies occur scattered throughout the tissues. Under the optical microscope these bodies show some internal structure and are crystalline. They are best seen in whole mounts of diseased mites which have been cleared and mounted in balsam or DePex medium. When viewed by polarized light the crystals become highly birefringent. The exact nature of these crystals, which have only been observed in diseased mites, is at present obscure. At first it was thought that they might be virus crystals, but examination of thin sections under the electron microscope showed that this was not so. The presence of these crystals is a valuable aid in diagnosing the disease.

Nothing is known at present of their chemical composition but they are extremely hard and are unaffected by Carnoy fixation, dehydration in ethanol, or clearing in xylol (Smith and Cressman, 1962).

The nucleic-acid composition of the virus has been investigated by Estes and Faust (1965). They found no DNA but estimated the RNA to be 10.6% of the dry weight of the initial virus preparation.

Panonychus ulmi (KOCH)

What appears to be a similar virus attacking the European red mite, *Panonychus ulmi* (Koch), has been recorded by Steinhaus (1959). The mites were collected from English walnuts near San Jose, California. Diseased individuals showed varying symptoms apparently depending on the age of the mite and the extent of the disease. The affected mites were obviously diseased, frequently becoming somewhat darker red in color, with blackened areas appearing upon the dorsal surface. They were commonly observed to undergo a trembling motion, and when turned over on their backs, were unable to right themselves. This is evidently a similar kind of paralysis to that which occurs in virus-affected citrus red mites. The diseased individuals of *P. ulmi* exuded fluid from the anal aperture causing them to adhere to the leaf in a manner similar to that displayed by *P. citri.*

Viruslike particles, approximately 40–60 mμ in diameter, were isolated from the diseased mites.

Experiments in the use of the first virus to control the citrus red mite are now in progress in Southern California. In the United Kingdom, the fruit-tree red spider which is probably the same as *P. ulmi,* has become a serious pest of apples owing to its having acquired a resistance to acaricides. Propagation of the virus in England might afford a much needed method of control of this pest.

Tetranychus SP. THE SPIDER MITE

A third possible mite virus is a transovarian factor which affects the morphology in spider mites. *Tetranychus* sp. It concerns the development of certain chemosensory setae on the legs. Experimental breeding suggests that a suppressor agent in the nonseta stock prevents the development of setae and that this agent is transmitted through ooplasm (Boudreaux, 1959).

REFERENCES

Boudreaux, H. B. (1959). A virus-like transovarian factor affecting morphology in spider mites. *J. Insect Pathol.* **1**, 270–280.

Estes, Z. E., and Faust, R. M. (1965). The nucleic-acid composition of a virus affecting the citrus red mite, *Panonychus citri* (McG.). *J. Invert. Pathol.* **7**, 259.

Smith, K. M., and Cressman, A. W. (1962). Birefringent crystals in virus-diseased citrus red mites. *J. Insect Pathol.* **4**, 229–236.

Smith, K. M., Hills, G. J., Munger, F., and Gilmore, J. (1959). A suspected virus disease of the citrus red mite *Panonychus citri* (McG.). *Nature* **184**, 70.

Steinhaus, E. A. (1959). Possible virus disease in European red mite. *J. Insect Pathol.* **1**, 435–437.

Author Index

A

Abul-Nasr, S., 238, *238*
Adams, J. R., 107, *107*
Aizawa, K., 14, 15, 25, 37, 42, 52, 55, 57, *58*, 112, *125*, 128, 134, 136, 142, *142, 143*, 172, 179, *191*
Allard, H. A., 198, 227
Amargier, A., 9, 19, *39*, 189, *194*
Anderson, C. W., 210, 227
Anderson, E. S., 78, *107*
Anderson, T. F., 80, *107*
Armstrong, J. A., 78, *107*
Arny, D. C., 210, 229
Aruga, H., 9, 16, 37, 42, 46, 57, 131, 135, 138, *143*, 150, 151, 152, *153, 154*, 173, 174, 178, *191*
Asayama, T., 175, *193*
Ayudaya, I. N., 135, 138, *143*
Ayuzawa, C., 112, *125*

B

Badami, R. S., 210, 227
Bailey, L., 89, 90, 91, 93, 94, 95, 96, 98, 99, *107*, 129, 130, 133, *143*, 150, *153*
Balch, R. E., 9, 20, 37
Baur, E., 3, *6*
Bawden, F. C., 180, *191*
Beckman, H. F., *191*
Beijerinck, M. W., 3, *6*
Bellett, A. J. D., 88, 107, 124, 125, *125*, *126*, 169, *170*

Bennett, C. W., 150, *153*
Ben Shaked, Y., 48, *57*, 131, *143*, 186, 192
Benz, G., 14, 18, 22, 37
Bergoin, M., 158, 159, 165, 166, 169, *171*
Bergold, G. H., 5, *6*, 10, 14, 20, 21, 22, 23, 25, 28, 37, *39*, 44, *57*, 59, 60, *62*, 66, 72, 110, *126*, 134, *143*, 181, *193*
Biliotti, E., 128, *239*
Bird, F. T., 9, 20, 37, 64, 72, 110, 111, *126*, 128, 134, *143*, 175, 177, 178, *191, 192*, 233, *239*
Björling, K., 209, 227
Black, L. M., 158, *170*, 217, 218, 219, *227, 228, 230*
Bockstahler, H. W., 225, *228*
Bolle, J., 5, *6*, 14, 37, 130, *143*
Boudreaux, H. B., 242, *242*
Brachet, J., 25, *38*
Bradley, R. H. F., 204, 205, *228*
Brakke, M. K., 49, *57*, 105, *107*, 218, *228*
Brazzel, J. R., 186, *192*
Brčák, J., 95, 98, *108*, 199, *228*
Breindl, V., 5, *7*, 22, 25, 37, *38*
Brenner, S., 53, *57*, 82, *108*
Brown, R. M., 67, 70, 71, *73*, 118, *126*
Bruckart, S. M., *191*
Bruehl, G. W., 211, *230*
Burgerjon, A., 233, 235, *239*
Burnside, C. E., 90, *108*, 133, *143*

C

Caspar, D. L. D., 53, 58, 114, 126
Chamberlin, F. S., 233, 239
Chambers, H., 186, 192
Chang, G. Y., 138 145, 174, 194
Chapman, A. J., 236, 238, 239
Chaudhuri, R. P., 207, 228
Clark, T. B., 10, 35, 37, 38, 87, 108
Clinch, P., 212, 228
Common, I. F. B., 11, 38
Conte, A., 130, 143
Coons, G. H., 225, 228
Cornalia, E., 5, 6
Cornuet, P., 198, 228
Costa, A. S., 197, 228
Cressman, A. W., 223, 230, 241, 242
Crick, F. H. C., 53, 57
Croissant, O., 106, 109, 128, 145
Cunningham, I., 157, 171

D

Dasgupta, B., 35, 38
David, W. A. L., 177, 178, 183, 184, 185, 192
Day, M. F., 11, 23, 38, 86, 87, 88, 108, 128, 133, 143, 148, 153, 156, 170, 182, 192, 203, 204, 206, 207, 228
DeBach, P., 238, 239
deJong, D. J., 10, 22, 39, 136, 144
de Meesters-Manger Cats, V., 199, 230
de Silva, D. M., 197, 228
Dewey, S., 225, 228
Dineen, J. P., 44, 58, 151, 152, 154
Dobrovolskaya, G. M., 28, 38
Duffus, J. E., 197, 207, 208, 228
Dulbecco, R., 156, 170
Dunn, P. H., 62, 63, 64, 73
Dutky, S. R., 12, 38, 233, 239

E

Eagle, H., 162, 170
Earle, W. R., 163, 170
Easterbrook, K. B., 169, 170
Eastop, V. F., 203, 228
Ehrhardt, P., 208, 223, 228
Elgee, D. E., 142, 144, 177, 192

Elmer, O. H., 198, 228
Elmore, J. C., 141, 143, 234, 238, 239
Escherich, K., 128, 143
Estes, Z. E., 241, 242

F

Falcon, L. A., 237, 238, 239
Farrant, J. L., 11, 23, 38, 128, 143
Faulkner, P., 15, 25, 38
Faust, R. M., 241, 242
Fort, S. W., 186, 194
Fraenkel-Conrat, H., 28, 38
Franklin, R. E., 25, 38, 114, 126
Franz, J., 142, 143
Frey, S., 98, 108
Frist, R. H., 115, 117, 121, 127, 140, 145
Frosch, F., 2, 3, 7
Fukuda, S., 9, 16, 37, 173, 178, 191
Fukushi, T., 218, 228
Furgala, B., 94, 95, 97, 108
Furuta, Y., 42, 57, 179, 191

G

Gamez, R., 226, 228
Gardiner, B. O. C., 177, 178, 183, 184, 185, 192
Gaw Zan-Yin, 168, 170
Gershenson, S. M., 10, 28, 38, 135, 143, 179, 189, 192
Getzin, L. W., 186, 192, 234, 239
Gibbs, A. J., 90, 93, 95, 96, 98, 99, 107, 129, 130, 133, 143, 150, 153
Gierer, A., 28, 38
Gilmore, J., 240, 242
Glaser, R. W., 177, 192
Glass, E. H., 233, 239
Goldschmidt, R., 156, 170
Gonzales, C. G., 226, 228
Grace, T. D. C., 6, 7, 102, 103, 108, 156, 157, 158, 162, 164, 165, 168, 169, 170
Gratia, A., 25, 38
Griggs, W. H., 226, 228
Grison, P., 233, 235, 238, 239
Gyrisco, G. C., 206, 229

H

Hale, R. L., 187, *193*
Hall, C. E., 11, *38*
Hall, I. M., 62, 63, 64, 73, 234, 238, *239*
Hamm, J. J., 63, 65, 72, 190, *192*
Hamman, P. J., 186, *192*
Hannah, A. E., 211, *230*
Harpaz, I., 48, *57*, 131, *143*, 186, *192*
Hashimoto, Y., 173, 178, *191*
Heidenreich, E., 100, *108*
Heinze, K., 207, 214, *228*
Hervey, G. E. R., 233, 238, *239*
Hills, G. J., 23, *39*, 44, 49, 52, 53, 56, *58*, 68, *73*, 75, 80, 82, 98, *108*, *109*, 114, 115, 122, 124, *126*, 129, 134, 137, 140, *145*, 181, *194*, 240, *242*
Hirumi, H., 158, 165, 166, 167, *170*, 218, 220, *229*
Hoggan, I. A., 198, *228*
Horie, J., 187, *193*
Horne, R. W., 53, *57*, 82, *108*, 122, *126*
Hosaka, Y., 52, 55, *58*
Houston, B. R., 211, *229*
Howland, A. F., 141, *143*
Huger, A., 61, 65, 68, 72, 100, 101, 107, *108*, 140, *143*, 189, *192*
Hughes, K. M., 62, 63, 64, 65, 66, 72, *73*
Hukuhara, T., 130, 135, 138, *143*, *144*, 148, 150, 151, *153*, 173, 174, 178, *191*
Hurpin, B., 100, 101, 105, *108*
Hutchinson, P. B., 214, *228*

I

Ignoffo, C. M., 12, *38*, 183, 184, 187, *192*, 236, 237, 238, *239*
Iida, S., 15, 25, *37*
Inman, D., 88, *107*
Irzykiewicz, H., 204, *228*
Ishikawa, Y., 175, *193*
Ishimori, J., 41, *58*
Israngkul, A., 42, *57*
Ito, T., 187, *193*
Ivanovsky, D., 2, 3, 7

J

Jaques, R. P., 128, 129, *144*, 151, *153*
Jeenor, R., 25, *38*
Jensen, D. D., 201, 222, 223, 224, 225, 226, *228*, *229*
Jírovec, O., 25, *37*
Jones, B. M., 157, *171*
Joshitake, N., 9, 16, *37*

K

Kaesberg, P., 52, *58*, 80, *108*
Kassanis, B., 172, *193*, 212, 213, 224, *228*
Kawai, T., 174, *191*
Kawase, S., 14, 26, 27, *38*, 46, 57, *58*
Kellen, W. R., 10, 35, 37, *38*, 87, *108*
Kelsey, J. M., 238, *239*
Kennedy, J. S., 203, *228*
Kikumoto, T., 199, 201, *228*
Kimura, I., 218, *228*
Kiso, A., 223, *230*
Kitajima, E., 150, *153*
Klug, A., 25, *38*, 53, *58*, 114, *126*
Kok, I. P., 28, *38*
Komárek, J., 5, 7, 22, *38*
Králik, O., 95, 98, *108*
Krczal, H., 225, *230*
Krieg, A., 23, 25, *39*, *65*, 72, 100, 101, 107, *108*, 112, 114, *126*, 134, 142, *143*, 144, 152, *153*
Krywienczyk, J., 181, 189, *193*
Kunkel, L. O., 215, 216, *228*
Kurisu, I., 151, *153*
Kvičala, B., 212, *229*

L

Langenbuch, R., 142, *143*
Laudeho, Y., 9, 19, *39*
Lee, P. E., 94, 95, 97, *108*, 223, *229*
Legg, J. T., 210, *229*
Leutenegger, B., 86, 98, *109*, 122, *126*
L'Heritier, P., 104, *108*
Lindegren, J. E., 10, 35, 37, *38*
Littau, V. C., 219, 222, *229*
Liu Nien Tsui, 168, *170*

Loeffler, F., 2, 3, 7
Lomakina, L. Ya., 222, *229*
Lotmar, R., 41, *58*
Loughheed, T. C., 157, *171*
Loughnane, J. B., 212, *228*
Lower, H. F., 61, 72
Lum, P. T. M., 87, *108*
Lwoff, A., 4, 7, 147, *154*

M

McEwen, F. L., 233, 238, *239*
MacGregor, D. R., 181, *193*
Maestri, A., 5, 7
Maramorosch, K., 134, *144*, 157, 158, 165, 166, *170, 171*, 201, 203, 218, 219, 220, 222, 224, 225, *229*
Markham, R., 98, *108*, 121, *126*
Martignoni, M. E., 65, 66, 72, 134, 141, *144*, 157, 158, 161, 162, 168, *171*, 176, 177, 178, *193*
Martin, D. F., 236, 238, *239*
Martouret, D., 238, *239*
Masera, E., 128, *144*
Matsui, C., 199, 201, *228*
Matthews, R. E. F., 214, *228*
Mercer, E. H., 86, 87, 88, 102, 103, *108*, 124, *126*, 169, *170*, 182, *192*
Merrill, H. M., 218, *229*
Milstead, J. E., 141, *144*
Mitsuhashi, J., 158, *171*, 218, 220, *229*
Miyajima, M., 128, *143*
Moore, D. H., 14, *39*
Morand, J. D., 198, *228*
Morgan, C., 14, 22, *39*
Morison, G. D., 94, *108*
Morris, O. N., 16, 19, 22, *39*
Mueller, W. C., 211, 214, *229*
Mulligan, T., 211, *230*
Munger F., 240, *242*
Murphy, P., 212, *228*

N

Nagashima, E., 131, *143*, 174, *191*
Nagington, J., 122, *126*
Nakamura, K., 179, *191*
Nault, L. R., 206, *229*
Neilson, M. M., 140, 142, *144*, 176, *193*

Newton, W., 198, *229*
Nishi, H., 201, *229*
Niven, J. S. F., 78, *107*

O

Olitsky, P. K., 198, *229*
Omura, H., 176, *194*
Orenski, S. W., 218, 220, *229*
Orlando, A., 4, 7
Orlob, G. B., 199, 200, 210, *229*
Osborn, H. T., 207, *229*
Ossiannilson, F., 199, 209, 227, *229*
Ossowski, L. L. J., 177, *193*, 238, *239*
Oswald, J. W., 211, *229*
Otomo, N., 176, *194*
Owada, M., 151, *153*

P

Paillot, A., 59, 65, 66, 72, 73, 111, *126*
Paine, J., 210, *229*
Pang, E.-Wa., 211, *229*
Paschke, J. D., 63, 65, 72
Pautard, F., *126*
Pelling, S., 18, *39*
Peters, D., 208, *230*
Pister, L., 25, *37*
Ponsen, M. B., 10, 22, *39*, 136, *144*, 207, 214, *230*
Potter, G., 11, 23, *38*, 128, *143*
Prentice, I. W., 207, 208, *229*
Proeseler, G., 225, *229*

R

Raun, E. S., 107, *108*
Ray, H. N., 35, *38*
Razvyazkina, G. M., 222, *229*
Reddy, D. V. R., 219, *230*
Reed, W., 3, 7
Reiner, C. E., 131, 141, *145*, 148, 153, *154*
Reiser, R., 183, *191, 194*
Rennie, J., 30, *39*
Richards, W. R., 211, *230*
Richardson, C. D., 186, *194*
Ripper, M., 152, *154*

Rivers, C. F., 63, 65, 73, 75, 98, *109*, 129, 132, 134, 137, 140, *144, 145,* 168, *171,* 177, 181, *193, 194*
Roberts, F. M., 203, 204, *230*
Rochow, W. F., 206, 209, 211, 214, 215, 229
Roegner-Aust, S., 147, *154*
Rose, H. M., 14, 22, *39*

S

Sato, M., 112, *127*
Scallion, R. J., 161, 162, 168, *171*
Schmid, P., 177, 178, *193*
Schmidt, H. B., 208, 222, *229*
Schmutterer, H., 196, *229*
Schneider, C. L., 225, 226, *228*
Schramm, G., 14, 28, 37, *38*
Sekiya, I., 222, *230*
Shikata, E., 218, 220, *228, 229*
Shinkai, A., 217, *229*
Shorey, H. H., 187, *193*
Shubnikova, E. A., 222, *229*
Sidor, C., 41, *58,* 137, *144,* 177, *193*
Silberschmidt, K., 4, 7
Simons, J. N., 207, 209, 210, *230*
Sinha, R. C., 219, *230*
Skuratovskaia, I. N., 28, *38*
Slykhuis, J. T., 211, *230*
Smirnoff, W. A., 18, *39,* 131, 136, 141, *144,* 150, *154,* 188, *193*
Smith, J. D., 28, *39*
Smith, K. M., 9, 10, 16, 18, 23, 30, 32, 34, *39,* 41, 44, 49, 52, 53, 56, *58,* 63, 65, 67, 68, 70, 71, *73,* 75, 79, 80, 82, 83, 84, 98, *108, 109,* 111, 112, 113, 114, 115, 117, 118, 121, 122, 124, *126, 127,* 129, 132, 134, 137, 139, 140, *144, 145,* 148, *154,* 168, *171,* 172, 176, 181, 188, *193, 194,* 199, 207, 212, 218, 223, 224, *230,* 240, 241, 242
Smith, O. J., 62, 63, 64, 73
Stairs, G. R., 35, *39,* 61, 68, *73,* 136, *145,* 180, *194*
Stanley, W. M., 4, 7
Stegwee, D., 207, 208, 214, *230*

Steinhaus, E. A., 14, *39,* 44, *58,* 59, 61, 63, *73,* 86, 95, 98, 102, *109,* 136, 142, *145,* 150, 151, 152, *154,* 172, *194,* 233, 235, 236, *239,* 241, *242*
Stelzer, M. J., 235, *239*
Stent, G. H., 30, *39*
Stephens, J. M., 177, *194*
Stubbs, L. L., 209, *230*
Sukhov, K. S., 199, *222, 230*
Surany, P., 100, 102, *109*
Suter, J., 44, *57*
Svoboda, J., 95, 98, *108*
Sylvester, E. S., 203, 204, 205, 206, 213, *230*

T

Takahashi, Y., 222, *230*
Tanabe, A. M., 131, 141, *145,* 148, 153, *154,* 179, *194*
Tanada, Y., 42, *58,* 61, 63, 65, 66, 68, *73,* 131, 138, 141, *145,* 148, 151, 152, 153, *154,* 174, 175, 178, 179, *194*
Tanaka, S., 151, *154*
Tarasevich, L. M., 14, *40*
Teissier, G., 104, *108*
TenBroeck, C., 218, *229*
Thomas, R. S., 82, 84, 89, *109*
Thompson, C. G., 63, 65, 66, 72, *73,* 142, *145,* 178, *194,* 233, 236, *239*
Toko, H. V., 207, 211, *230*
Trager, W., 157, 160, 168, *171*
Trontl, Z. M., 115, 117, 121, *127,* 140, *145*

V

Vago, C., 9, *40,* 95, 100, 101, 105, 106, *108, 109,* 128, *145,* 152, *154,* 158, 159, 165, 166, 169, *171,* 174, 189, *194*
van der Want, J. P. H., 204, *230*
Vanderzant, E. S., 183, 186, *194*
van Hoof, H. A., 204, *230*
van Soest, W., 199, *230*
Vitas, K. I., 28, *38*
von Prowazek, S., 5, 7
Volk, von J., 225, *230*

W

Wagner, R. P., 158, *171*
Walters, H. J., 196, 198, 199, *230*
Wasser, H. B., 61, *73*, 102, *109*
Watanabe, H., 42, 46, *57*, 151, *153*, 173, 174, *191*
Waterson, A. P., 180, *194*
Watson, J. D., 53, *57*
Watson, M. A., 203, 204, 211, 213, 226, *228*, *230*
Weiser, J., 106, *109*
Wellington, E. F., 14, 25, 26, 28, 37, *40*, 62, 71, *73*
Whalen, M. M., 20, *37*, 134, *143*
Whitcomb, R. F., 218, *230*
White, G. F., 130, *145*
Wilcox, T. A., 107, *107*
Williams, R. C., 79, 80, 82, 83, 84, *109*
Wilson, F., 129, *145*
Wittig, G., 44, *58*, 65, *73*
Wolfenbarger, D. A., 235, *239*
Woods, R. D., 90, 93, 95, 96, 98, 99, *107*, 129, 130, 133, *143*, 150, *153*
Woolcombe, T. M., 207, 208, 229

Wyatt, G. R., 13, 25, 28, *39*, *40*, 71, *73*, 157, *171*
Wyatt, S. S., 157, 160, *171*
Wyckoff, R. W. G., 41, *58*, 112, *126*, 132, *145*, 172, 188, *193*

X

Xeros, N., 9, 16, 18, 19, 30, 32, *39*, *40*, 41, 42, 49, 57, *58*, 65, *73*, 75, *109*, 111, 112, 113, 114, 122, 124, *126*, *127*, 172, 188, *193*, *194*

Y

Yamafuji, K., 112, *127*, 176, *194*
Yoshihara, F., 112, *127*
Yoshii, H., 223, *230*
Yoshitake, N., 42, 46, *57*, 135, 138, *143*, 151, 152, *153*, 173, 174, *191*

Z

Zia Tien un, 168, *170*
Zillinsky, R. J., 211, *230*
Zitcer, E. M., 158, *171*
Zlotkin, E., 48, *57*

Subject Index

A

Abutilon sp., 3, 4
Abraxas grossulariata, 16
Acacia mollissima Willd., 238
Acetic acid, 151, 190
Acridine orange, 124
Acrobasis sp., 106
Aëdes aegypti (L), 87, 218
Aëdes sierrensis Ludlow, 37
Aëdes taeniorhynchus (Wiedemann), 87
Agallia constricta, 218, 220, 221
Agalliopsis novella (Say.), 217
Aglais urticae (Linn.), 136
Agriotes obscurus (Linn.), 75, 98
Agrotis segetum Schiff, 59
Albuminoid spheres, 107
Aleyrodidae, 199
Alfalfa caterpillar, 152, 153
Alpine tent caterpillar, 18
Alsophila pometaria (Harris), 136, 176
Amidoschwartz stain, 189
Amino acids, 25–28, 57, 71, 89, 160, 162, 164
Amphorophora lactucae (Linn.), 208
Angleshades moth, 48
Aniline methyl blue, 189
Anopheles freeborni Aitken, 37
Anopheles subpictus Grassi, 35
Anosia plexippus (Linn.), 165
Antheraea eucalypti Scott, 124, 162, 165, 169
  virus from, 102
    host range, 102
    isolation of, 103
    morphology and ultrastructure, 104
    symptomatology and pathology, 103
Antheraea mylitta Drury, 52, 53, 80
Antheraea pernyi, Guérin-Ménéville, 10, 28, 29, 42, 44, 45, 50–54, 138, 169
Apanteles sp., 142
Aphis gossypii Glover, 209, 210
Aphis nasturtii (Kalt.), 210
Aphid–virus relationships, 198, 199, 201, 203
  artificial feeding, 213
  circulative viruses, 206
  propagative viruses, 207
  stylet-borne viruses, 203
  variations in transmission, 209
  virus injection, 213
Apis mellifera (Linn.), 89, 95
Aporia crataegi (Linn.), 25
Arachnida, 240
Arboviruses, 3
Arctia caja Linn, 41, 52, 138
Arctia villica Linn, 41, 44
Arginine, 14, 28
Argyrotaenia velutinana (Walker), 61, 233
Armyworms, 61, 92, 102, 114, 141, 160
Arsenic acid, 151
Artificial feeding
  aphids, 213
  larvae
    media, 182
    rearing containers, 184

Arthropoda, 4, 240
Aster leafhopper, 166
Aster yellows, 172, 178, 215, 216, 226
Aulacorthum solani (Kalt.), 198
Autographa brassicae Riley, 55, 56
Autographa californica Speyer, 172
Automeris io, 53
Azocarmine, 190

B

Bacillus popilliae Dutky, 231
Bacillus thuringiensis Berliner, 231
Barathra brassicae (Linn.), 196
Barley yellow dwarf, 209–211
Bee paralysis virus
  acute, 89, 129, 133
    host range, 90
    isolation of, 90
    morphology, 92, 93
    symptomatology and pathology, 90, 91
  chronic 94, 129, 133
    host range, 94
    isolation of, 94
    morphology, 95, 96
    symptomatology and pathology, 94, 91
Beet leaf curl, 225
Beet "savoy," 225
Beet yellows, 209
Bibio marci (Linn.), 75, 82
Bioassay, 187
Biological control, 231
  application of virus, 234
  examples of virus control, 238
  preparing virus, 235
  selection of viruses, 232
  standardization of virus preparations, 236
  variable factors, 232
Biological transmission (of plant viruses), 215
Blood cells, 32–34
Bollworm, 235
Bombus agrorum (Fabr.), 90
Bombus hortorum (Linn.), 90
Bombus lucorum (Linn.), 90

Bombus ruderaris (Muller), 90
Bombus terrestris (Linn.), 90
Bombyx mori L., 10, 14, 24–27, 29, 42, 44, 52, 55, 71, 75, 111, 112, 128, 131, 136, 147, 159, 160, 165, 169, 174, 175
"Boules hyalines," 66, 67
Bragg reflections, 80
Brimstone butterfly, 138
"Broken" tulips, 2
Bromophenol blue, 188
Buck eye caterpillar, 63
Buffalo black, 189
Buff-tip moth, 137
Bumblebees, 90
Bupalus piniarius (Linn.), 152, 168

C

Cabbage black ringspot, 212
Cabbage leaf powder, 184
Cabbage looper, 12, 63, 65, 178, 187, 232–235
Cabbage worms, 147, 232
Cacoecia murinana (Hübner), 71
Calliphora vomitoria (Linn.), 75
Callistephus chinensis, 225
Calophasia lunula (Hufnagel), 52
Capsid, 4
Capsomere, 4
Capsules (granules), 60, 61, 66, 118, 119
  chemical composition, 62
  physicochemical properties, 62
  shape, size, and structure, 61
Carbolfuchsin, 189
Carbon dioxide sensitivity, 104
Carbon tetrachloride, 92
Carpocapsa pomonella L. 61, 63, 68
Carrier virus, 213
Cartwheel forms, 124
Cauliflower mosaic, 212
Celery yellows, 223
Cell culture, 155
Cerura vinula (Linn.), 148
Cesium chloride, 105
Chick embryos, 179
Choristoneura fumiferana (Clemens), 64, 68, 71, 148, 175, 179

Choristoneura murinana Hübner, 59, 71
Chrysopa perla L. 10
Chrysopa spp., 41, 139
Cicadellidae, 158
Circulative (viruses), 206, 209
Cirphis unipuncta Haworth, 92
Cirphis, unipuncto, virus from, 102
Citrus red mite, 240
Clover club leaf, 217
Cold treatment, 131
Colias eurytheme Boisd., 112, 131, 136, 138, 141, 148, 152, 173
Colias lesbia Fabr., 136
Colladonus montanus Van Duzee, 223, 224
Color break, 2
Comma butterfly, 137
Corn earworm, 178
Corpora pedunculata, 90, 91
Crystalline lattice, 11, 12
Cucumber mosaic virus, 198
Culex sp., 87
Culex tarsalis Coq., 10
  polyhedra, 36
  possible nuclear polyhedrosis in, 35
  symptomatology and pathology, 36
  transmission, 36
  virus, 37
Cutworms, 59, 63
Cytocidal virus, 147
Cytopathic effect (CPE), 155, 168
Cytoplasmic polyhedra, 42–47, 50, 51
Cytoplasmic polyhedroses, 20, 26, 27, 41, 44, 45

D

Dalbulus maidis (DeLong), 225
Dandelion yellow mosaic, 213
Dasychira pudibunda (Linn.), 44, 152
Delphacodes striatella (Fallen), 217, 222
Delphocephalus striatus L., 222
Deltocephalus (*Imazuma*) *dorsalis* Mots., 217
Dendrolimus spectabilis (Butler), 173, 178
Density gradient centrifugation, 49, 51
Deoxyribonucleic acid (DNA), 83

Deoxyribonucleic acid, isolation of, 28
Diprion hercyniae (Hartig), 19
Diprion sp., 9
Double shadowing, 80, 82, 83
Drosophila funebris, 104
Drosophila gibberosa, 104
Drosophila melanogaster, 104
Drosophila prosaltans, 104
Drosophila σ virus, 104
  host range, 104
  symptomatology, 104
  virus, 105
Drosophila simulans, 104
Drosophila wellistoni, 104

E

Earwigs, 197
EDTA, 151
Epitrix cucumeris, Harr., 199
Equine encephalomyelitis, 218
Erannis tiliaria (Harris), 18, 136
Erithacus rubecula (Linn.), 142
Escherichia coli Castellani and Chalmers, 78
Estigmene acrea (Drury), 42, 43, 48, 172
Euchlaena mexicana, 225
Eucosma griseana, 64
European corn borer, 107
European pine sawfly, 19
European spruce sawfly, 19
Euxoa segetum (Schiff.), 172

F

Fibroblasts, 158, 159
Fig mosaic, 227
Fluorescent staining, 78, 189
Fluorocarbon, 20
"Flying pin," 197
Foot-and-mouth disease, 2
Fragaria vesca Linn., 198
Freeze-drying, 80
Fumarate, 157

G

Galleria mellonella Linn., 6, 35, 86–88, 122, 133, 136, 148, 150, 176, 179
Geisha distinctissima Wal., 223

Giemsa solution, 188, 189
Gilpinia hercyniae (Htg.), 9
Gipsy moth, 16, 28, 177
Gonepteryx rhammi (Linn.), 75, 138
Gradocol membranes, 105
Granuloses, 59, 117
   chemistry of virus of, 71
   isolation of virus of, 66
   long virus threads associated with, 70
   morphology and ultrastructure of virus
      from, 68
   replication of virus, 115
Grasserie, 5
Grasshoppers, 196, 197
Granules, *see* Capsules
Group specificity, 135
Greater wax moth, 86, 133, 136, 148

H

Harrisina brillans B. & McD., 62, 63
Heidenhain's hematoxylin, 190
Heidenreich's disease, 100, 102
HeLa cells, 156
Heliothis virescens Fabr., 235, 238
Heliothis zea (Boddie), 186, 235, 237, 238
Helix, 114
Helix aspersa Müller, 158
"Helper" viruses, 212, 213
Hemalum, 189
Hematoxylin, 103
Hemerobius spp., 41, 137, 139
Hemerobius stigma Steph., 12
Hemocytometer, 233
Henbrane, 226
Hepialus humuli (Linn.), 75
Hepialus lupulinus (Linn.), 75
Histolytic disease, 100
Hop mosaic, 210
Hyadaphis brassicae, 214
Hyloicus pinastri, 47
Hyoscyamus mosaic, 199, 201, 203, 226
Hyphantria cunea, 174

I

Icosahedra, 52, 53, 80–84, 240
Immunity, 175

Inapparent infection, 146
Inazuma dorsalis, 217, 224
Incubation period, 215
Infection
   inapparent, 146
   latent, 146
   per os, 128
   occult virus, 146
   subclinical, 146
Infectious variegation, 3
Infective units (IU), 125
Infective virus, 147
Insect virus diseases, different kinds of, 8
Interference (between viruses), 172
Intimate membrane, 22–24, 68

J

Jackpine sawfly, 19
Japanese beetle, 231
Jassidae, 195
Jaundice, 5, 16
Junonia coenia Hübner, 63

K

Kapselvirus-Krankheit, 60
Ketoglutarate, 157
Kotochalia junodi (Heyl.), 177

L

Lace bug, 225
Lambdina fiscellaria somniaria Hulst., 16, 22
Laothoe populi Linn., 57
Latency, definition of, 146
Latent infections, 146
   conditions governing, 150
   examples of, 148
   induction of, 150
   state of virus in, 147
Latent period, 111, 207
Leafhoppers, 166, 203, 215–217, 219, 220–223, 225
Leucania unipuncta, 160
"Life-cycle," 110
Liriomyza langei Frick, 197
Lymantria monacha L., 16, 22, 128
Lymantria (*Porthetria*) *dispar*, 16, 75, 158, 159, 165, 169

M

Maclura aurantiaca Nutt., 152
Macrosiphum euphorbiae Thomas, 198
Macrosiphum geranicola (Hille Ris Lambers), 209
Macrosiphum granarium (Kirby), 211
Macrosiphum pisi Kalt., 210
Macrosteles fascifrons Stal., 215, 218
Malacosoma alpicola (Standinger), 18, 136
Malacosoma americanum (Fabr.), 136
Malacosoma disstria (Hübner), 134, 136
Malacosoma fragile (Stretch), 235
Malacosoma pluviale Dyar, 136
Malacosoma spp., 136
"Maladie transparente," 100, 102
Malate, 157
Malaya disease, 101
Mechanical vectors (of plant viruses), 196
Melanoplus existientialis, 196
Melolontha hippocastani Fabr., 100
Melolontha melolontha (Linn.), viruses from, 100, 105
host range, 100, 106
isolation of, 101, 106
morphology and ultrastructure, 101, 106
symptomatology and pathology, 100, 106
Melolontha sp., 75, 98, 100
Methanyl yellow, 189
Microcrystals, 76, 79, 218
Microinjection, 218
Microsporidia, 152
Microvesicles, 65
Milky disease, 231
Mixed infections, 172
Moderate virus, 147
Molecular crystalline lattice, 10–12, 60
"Molecular sickness," 117, 118
Monarch butterfly, 165, 167
Monoiodoacetic acid, 151
Mottle virus, 207, 212
Mulberry leaf powder, 186, 187

"Mushroom bodies," 90, 91
Mycetophilus sp., 75
Myzus ascalonicus, 210
Myzus circumflexus (Buckt.), 198
Myzus persicae Sulz., 199, 202, 207–212, 222, 223

N

Natada nararia (Moore), 65
Negative "staining," 23
Neodiprion pratti banksianae Rohiver, 19
Neodiprion sertifer (Geoffroy), 19, 112, 133, 134, 141, 152, 238
Neodiprion sp., 9
Neodiprion swainei, 131, 141
Nephotettix apicalis var. *cincticeps*, 215, 217, 218, 224
Neuroptera, 9, 10, 12, 41, 137
Nicotiana clevelandii, 213
Nicotiana glutinosa, 196
Noninclusion-body viruses, 8, 9, 74
Nonpersistent (viruses), 203, 206
Nuclear net, 112
Nuclear polyhedra, 10
amino acid composition of protein, 14, 15
Nuclear polyhedrosis, 9
Nuclear polyhedrosis viruses, 9, 111
latent period of, 111
morphology of virus replication, 112
multiplication of, 112
Nun moth, 16
Nymphalis antiopa (Linn.), 176
Nymphalis io Linn., 136, 181

O

Occult virus, 131, 146
Operophtera brumata (Linn.), 148, 152
Orange dwarf disease, 223
Orthosia incerta (Hufnagel), 22
Oryctes rhinoceros (Linn.), 101
Ostrinia nubilalis Hübner, suspected virus from, 107
Ourapteryx sambucaria L., 42, 49
Ovarian tissue, 157, 159
Ovariole, 157, 158

P

Paleacrita vernata (Peck), 176
Panaxia dominula, 10, 15, 16
Panonychus citri (McG), 223, 240
Panonychus ulmi (Koch), 241
Paper chromatography, 26
Parafilm, 214
Pea enation mosaic (PEMV), 206
Pear decline, 226
Pear psyllid, 226
Penicillin, 75, 165
Pentatrichopus fragaefolii Cock., 198
Peridroma margaritosa Haw., 59
Peridroma saucia (Hübner), 158, 161, 176
Peridroma sp., 151
Persectania ewingii Wester., 61
Persistent (viruses), 203, 206
Phagocytosis, 34, 35
Phalera bucephala, 137
Phigalea titea Cramer, 136
Phlogophora meticulosa (Linn.), 46, 48
"Phony" disease (peach), 227
Phosphotungstic acid, 23, 53, 70, 71, 82, 190
Phorodon humuli (Schrank), 210
Photographic rotation technique, 98
Phryqanidia californica (Packard), 178
Phyllotreta spp., 197
Physalis floridana, 223
Pieris brassicae (Linn.), 59, 63–65, 69, 75, 82, 87, 129, 132, 133, 140, 148–150, 169, 174, 177, 181, 183–185, 236, 238
Pieris napi (Linn.), 75, 140
Pieris rapae (Linn.), 48, 59, 63, 64, 75, 117, 140, 147, 148, 172, 178, 238
Piesma cinerea, 225
Piesma quadrata, 225, 226
Pine caterpillar, 173
Pine looper, 148
Pink bollworm, 183
Plant hopper, 217
Plaque technique, 156
Plusia chalcytes (Esp.), 9, 19
Poliovirus, 155
Polygonia c-album (Linn.), 137, 138

Polyhedra, 4, 5
  icosahedral, 46
  hexagonal, 46, 130
  tetragonal, 130
Polyhedroses
  cytoplasmic type, 41
    polyhedra, 42
    chemical composition, 44
    physicochemical properties, 44
    size and structure, 42
    symptomatology and pathology, 47
    virus
      chemistry of, 57
      isolation of, 49
      morphology and ultrastructure of, 52
  nuclear type, 9
    polyhedra, 10
    chemical composition, 14
    physicochemical properties, 11
    size and structure, 10
    symptomatology and pathology, 15
    virus
      isolation of, 20
      chemistry of, 25
      morphology and ultrastructure of, 22
Popillia japonica Newn., 231
Poplar sawfly, 136
Porthetria dispar L., 10, 14, 16, 22, 25, 75, 134, 137, 139, 152, 181
Potato aucuba mosaic, 212
Potato flea beetle, 196, 199
Potato leafroll, 207, 208
Potato virus A, 212, 213
Potato virus X, 212
Potato virus Y, 209, 210, 212, 213
Potato yellow dwarf, 215
Prodenia litura Fabr., 131, 186, 238
Propagative (viruses), 203, 207
Propolyhedra, 16, 18
Provirus, 146
Psammotettix striatus Fall., 222
Pseudaletia unipuncta Haworth, 61, 65, 68, 138, 152, 172, 175
Pseudococcus citri (Risso), 198
Pseudococcus maritimus Ehr., 198, 199

Pseudograsserie, 59
Pseudorosette (of oats), 222
Psylla pyricola Foerster, 226
Psylliodes spp. 196
Pterolocera amplicornis Walker, 11
Puss moth, 148

R

Radish yellows, 207
Rearing jars, 184
Recurvaria milleri Busck, 181
Red-banded leaf roller, 233
Resistance, 175
Rhinoceris annulatus (Linn.), 142
Rhinoceros beetle, 101
Rhopalosiphum fitchii (Sanderson), 210
Rhopalosiphum maidis (Fitch), 211
Rhopalosiphum padi (Linn.), 211
Ribonuclease, 29
Ribonucleic acid (RNA), 28, 29, 53, 57
Rice dwarf disease, 215, 217, 218
Rice stripe disease, 217
Ring zone, 18, 19

S

Sabulodes caberata (Guenée), 172
Sacbrood virus, 95, 97
    host range, 95
    isolation of, 98
    morphology and ultrastructure, 98
    symptomatology and pathology, 95
Salivary sheath, 199, 202
Salt marsh catepillar, 48
Samia cynthia pryeri, 175
Sandal spike disease, 227
Sawflies, 9, 19, 112, 142
Scorzonera hispanica (Linn.), 152
Semipersistent (viruses), 203, 206
Sericesthis iridescent virus (SIV), 86, 118, 148, 180, 182
    amino acid analysis of, 89
    chemical and physical properties of, 88
    host range, 87
    isolation of, 87
    morphology and ultrastructure, 88
    replication of, 122
    symptomatology and pathology, 87

Sericesthis pruinosa (Dalman), 86, 98
Serine, 160
Serology of insect viruses, 180
"Spherical" viruses, 52
Silkworm viruses, 14
    amino acid contents of, 26, 27
Silkworms, 2, 14 42, 55, 57, 111, 128, 165, 173, 175
Sodium azide, 151
Sodium carbonate, 11, 21, 50, 52
Sodium cyanide, 151
Sodium fluoride, 151
Sodium hydroxide, 32
Sodium hypochlorite, 11
Sonchus oleraceus, 208
Sowbane mosaic, 197
Sowthistle yellow vein virus, 207, 208
Sphinx ligustri Linn., 134, 138, 188
Spider mites, 240
Spread, methods of virus, 128
Spruce budworm, 61
St. Mark's fly, 75
Staining methods for optical microscopy, 188
Strawberry crinkle virus, 208
Strawberry virus-3, 207
Streptomycin, 134, 165, 168
Stressors, 150
Stylet-borne (viruses), 203
Subclinical infection, 146
Submoderate virus, 147
Succinate, 157
Sucrose, 87, 157
Sugarbeet yellows, 206
Synergism, 172

T

Telea polyphemus Cram., 132
Tenebrio molitor (Linn.), 75, 98
Tent caterpillars, 18
Teosinte, 225
Tettigonia cantans, 196
Tettigonia viridissima, 196
Tetranychus sp., 241
T-Even bacteriophage, 78, 80
Thaumetopoea pityocampa (Schiff.) 42, 49, 114, 235
Thaumetopoea wilkinsoni Tams, 48

Tiger moth, 10, 16
Tineola bisselliella Hummel, 148
Tipula iridescent virus (TIV), 52, 55, 74
  chemistry of, 82
  host range, 75
  isolation of, 78
  morpholoyy and ultrastructure, 80
  optical properties of, 78
  replication of, 122
  symptomatology and pathology, 76
Tipula livida Van der Wulp, 75
Tipula oleracea (Linn.), 35, 75
Tipula paludosa, Meig., 9, 10, 21, 22,
    106, 122
  nuclear polyhedrosis of, 30, 31
Tipula spp., 75
Tissue culture, 155
  introduction, 155
  results achieved, 168
  techniques and media, 160
Tobacco budworm, 233
Tobacco mosaic virus, 23
  vector relationships of, 197
Tobacco rosette, 207
Toluidine blue, 103
Top component, 121
Toxoptera graminum Rond., 209
Transmission of viruses,
  artificial methods of, 133
  cross, 135
  transovarial, 130
  transovum, 130
Tree-top disease, 16
Trichiocampus irregularis (Dyar), 136
Trichiocampus viminalis Fall., 136
Trichoplusia ni Hübner, 65, 128, 151,
    178, 183, 233
Tropaeolum majus (Linn.), 177
Trypsin, 156, 158
Turnip flea beetle, 197
Turnip yellow mosaic, 197

U

Uranyl acetate, 124

V

Vanessa atalanta (Linn.), 136
Vanessa cardui (Linn.), 136, 139

Vaccinia, 155
Vein-distorting virus, 207
Virion, 4
Virogenic stroma, 113, 114
Virus
  cytocidal, 147
  infective, 147
  methods of spread of, 140
  moderate, 147
  mutations, 179
  occult, 146
  strains, 179
  submoderate, 147
  vegetative, 147
"Viruskapseln," 60
Virus rods, 13

W

Wassersucht virus, 98
  host range, 100
  isolation of, 101
  morphology, 101
  symptomatology and pathology, 100
Watery degeneration, 100
Wattle bagworm, 177, 238
Wax moth, 136, 176
Western Oak looper, 16
Western X-disease, 223
Whiteflies, 199
Willow sawfly, 136
Winter moth, 148, 152
Winter wheat mosaic, 222
Wound-tumor virus, 218–221

X

X-Bodies, 222
X-Rays, 10

Y

Yellow fever, 2, 3
Yellow net virus, 213

Z

Zea mays, 225
Ziehl fuchsin, 189